INTERNATIONAL ENERGY AGENCY

THINGS THAT GO BLIP IN THE NIGHT

Standby Power and How to Limit it

Benoit Lebot
Benoit.lebot@equipement.gouv.fr
benoitlebot@aol.com

INTERNATIONAL ENERGY AGENCY

9, rue de la Fédération,
75739 Paris Cedex 15, France

The International Energy Agency (IEA) is an autonomous body which was established in November 1974 within the framework of the Organisation for Economic Co-operation and Development (OECD) to implement an international energy programme.

It carries out a comprehensive programme of energy co-operation among twenty-five* of the OECD's thirty Member countries. The basic aims of the IEA are:

- To maintain and improve systems for coping with oil supply disruptions;
- To promote rational energy policies in a global context through co-operative relations with non-member countries, industry and international organisations;
- To operate a permanent information system on the international oil market;
- To improve the world's energy supply and demand structure by developing alternative energy sources and increasing the efficiency of energy use;
- To assist in the integration of environmental and energy policies.

** IEA Member countries: Australia, Austria, Belgium, Canada, the Czech Republic, Denmark, Finland, France, Germany, Greece, Hungary, Ireland, Italy, Japan, Luxembourg, the Netherlands, New Zealand, Norway, Portugal, Spain, Sweden, Switzerland, Turkey, the United Kingdom, the United States. The European Commission also takes part in the work of the IEA.*

ORGANISATION FOR ECONOMIC CO-OPERATION AND DEVELOPMENT

Pursuant to Article 1 of the Convention signed in Paris on 14th December 1960, and which came into force on 30th September 1961, the Organisation for Economic Co-operation and Development (OECD) shall promote policies designed:

- To achieve the highest sustainable economic growth and employment and a rising standard of living in Member countries, while maintaining financial stability, and thus to contribute to the development of the world economy;
- To contribute to sound economic expansion in Member as well as non-member countries in the process of economic development; and
- To contribute to the expansion of world trade on a multilateral, non-discriminatory basis in accordance with international obligations.

The original Member countries of the OECD are Austria, Belgium, Canada, Denmark, France, Germany, Greece, Iceland, Ireland, Italy, Luxembourg, the Netherlands, Norway, Portugal, Spain, Sweden, Switzerland, Turkey, the United Kingdom and the United States. The following countries became Members subsequently through accession at the dates indicated hereafter: Japan (28th April 1964), Finland (28th January 1969), Australia (7th June 1971), New Zealand (29th May 1973), Mexico (18th May 1994), the Czech Republic (21st December 1995), Hungary (7th May 1996), Poland (22nd November 1996), the Republic of Korea (12th December 1996) and Slovakia (28th September 2000). The Commission of the European Communities takes part in the work of the OECD (Article 13 of the OECD Convention).

PREFACE

I was amazed when I first learned the dimensions of standby power. In my own home, appliances that I thought were "off" were actually consuming considerable power. My home is hardly unusual. Taking all of the homes in IEA Member countries, standby power accounts for 10 per cent of residential electricity demand. I also learned that we can reduce standby power consumption by about 75 per cent using cost-effective technologies and design changes. It is a classic "low-hanging" fruit that can be plucked. We get energy savings and cost savings for consumers.

International collaboration can help our Member countries achieve these savings faster and cheaper than if countries acted independently. When we started this project, some countries or regions were developing programmes to address standby power waste while others were not aware of the problem. Governments were proceeding with different definitions and protocols. The IEA helped get countries talking to one another to co-ordinate their programmes. As a result, several countries have developed more effective programmes and others have a better understanding of what is needed.

Although obstacles remain, I am happy that the IEA has been able to make a tangible contribution to saving energy. With additional effort, best practice in reducing standby losses can be standard practice in less than ten years. With this book, we hope to spread our message more widely and provide some recommendations about how to bring this about.

Robert Priddle
Executive Director

ACKNOWLEDGEMENTS

The topic of standby power losses was brought to the attention of the energy efficiency community and to the IEA by Dr Alan Meier of Lawrence Berkeley National Laboratory in the United States in the mid-1990s. His groundbreaking work has been instrumental in advances made in the field of standby power losses.

In January 1999, the IEA Secretariat established a work programme to help IEA Member countries design policies and programmes to reduce standby power losses. Benoît Lebot, of the IEA, managed this programme, with the support of the Head of the Energy Efficiency Policy Analysis Division, first Jean-Pierre Des Rosiers and then Carmen Difiglio.

The programme has been a collaborative effort from the beginning. Many representatives of governments, industry and consumer groups participated in a series of workshops and task forces. These workshops, together with Internet discussion groups, have greatly facilitated international information exchange and collaboration.

This report draws on work by the task forces. Particular mention should be made of the input to this publication from Alan Meier and Karen Rosen of Lawrence Berkeley National Laboratory, United States; Ken Salaets of the Information Technology Industry Council, United States; Hans Paul Siderius of NOVEM, Netherlands; and Fiona Mullins of Environmental Resources Management, United Kingdom.

TABLE OF CONTENTS

FIGURES AND TABLES

EXECUTIVE SUMMARY

Standby power is a new phenomenon. Thirty years ago, many consumers could simply unplug their refrigerators and go on holiday, assured that their electricity meter would just stop. Today, however, meters continue to turn. Residences act as if they had a life of their own. Standby power consumption is the cause.

A growing number of electrical devices are designed to draw power 24 hours a day, seven days a week. Much of this energy may be consumed while the product is not being fully utilised by the consumer, while it is merely "standing by". For many products, such as printers and copiers, the standby or "sleep" mode provides an important and distinct benefit, decreasing significantly the total energy the device consumes. But not all standby uses are so benign. And recent field studies indicate that between 3 and 13 per cent of residential electricity use in OECD Member countries can be attributed to standby power consumption. That amount is projected to increase substantially with the growing number of electronic devices and the trend to home and office networks. More can and should be done to reduce standby power consumption.

Many IEA countries have launched programmes to reduce standby power consumption in the most used appliances, such as television sets and personal computers. However, these programmes should now be expanded to cover many other products.

Electricity consumption in standby modes is often far higher than necessary. For some products, existing engineering practices could greatly reduce standby power use at relatively low cost and without affecting how the product operates or consumer satisfaction. More widespread use of existing power management technology could reduce total standby energy consumption by as much as 75 per cent in some appliances. The corresponding reduction in carbon emissions could be a cost-effective component in an overall global strategy to reduce greenhouse gas emissions.

The IEA believes that, within ten years, products with optimised power management could be the norm rather than the exception. A window of opportunity is available now to ensure that this vision becomes a reality. By working with industry, consumers and other interested parties, IEA Member countries can encourage the design and introduction of new, more efficient appliances that meet the needs of both customers *and* the environment. Concerted international action to rein in standby power now could greatly reduce unnecessary future growth in standby power. International collaboration is essential to achieve this vision, since so many products and components are traded internationally. The problem will become more intractable once more networked products with high standby power requirements have been purchased for use in offices and homes.

THE EVOLUTION OF STANDBY POWER

INTRODUCTION

Many people are surprised to learn that they cannot stop their appliances from using electricity without pulling the plug from the socket or switching the power off at the wall. Today, televisions, video-cassette recorders (VCRs), mobile telephone chargers, computers, fax machines and many other appliances consume electricity 24 hours a day. These appliances are said to be in "standby mode" when they are not being used or are not performing their primary function. The power that appliances consume while they are in standby mode is called standby power.

In the late 1990s, the standby phenomenon was widely known as "leaking electricity" – a catching, but not quite accurate name. Along with the electronics industry generally, the IEA now refers to "leaking electricity" by the more neutral term "standby power".

Many appliances require a small level of electricity for standby functions – to power a built-in clock, respond to programming or respond to remote commands. In other devices, energy is simply wasted when the appliance is not doing anything at all.

The discussion in this book is limited to standby power losses from appliances that run on electricity. Standby energy consumption also occurs in gas fireplaces, gas refrigerators, oil boilers, gas and oil water heaters, gas clothes dryers and gas cooking equipment. Standby losses from these appliances are not analysed here

EVOLUTION

Thirty years ago, it was possible physically to disconnect most household appliances from the power source when they were not in use. Householders in many countries could unplug their refrigerators and go on vacation assured that their electricity meter would just stop.

Some of the first consumer goods to use standby power were audio and video products. Manufacturers equipped televisions, VCRs and stereo systems with standby functions such as remote control, time-keeping and memory. Next came clock displays on kitchen ranges and microwave ovens, remote control for garage doors and a variety of other functions that required continuous power use. Examples of products that have only recently begun to draw power continuously include fireplaces, ceiling fans, air-conditioners and white goods with electronic controls. Increasingly, appliances have external energy supplies ("wall packs"), which consume power even if the device they are supplying is not performing any operation. A typical modern home may have dozens of products that consume energy around the clock, even when they are not being used.

The development of "intelligent" energy management systems in homes and offices will require many more appliances to remain in standby mode in order to communicate with other devices or to receive data from external networks via satellite, microwave, telephone line or cable. Refrigerators, washing machines and other appliances will be able to communicate with computers and mobile phones. Multi-functional devices will serve as televisions, computers and video entertainment systems. Many products will require some continuous power to maintain connections and to be able to respond to requests from the network at any time. Appliances that traditionally did not need standby power will require it once they are networked. Appliances that already need standby power may require more to maintain connections. Standby power consumption will expand rapidly if the number of networked appliances increases.

Standby energy use now accounts for a significant share of total appliance energy use. In the residential sector in OECD countries, it is responsible for 1.5 per cent of those countries' total electricity consumption (128 TWh/yr; see Annex 2). It is estimated that standby power in the residential and commercial sectors combined may account for 2.2 per cent of total OECD electricity consumption and almost 1 per cent of total OECD carbon emissions. Much of this consumption is unnecessary, and significant savings could be achieved by reducing it. Studies in Germany show that standby power is responsible for 20 TWh of German electricity consumption and about 10 per cent of electricity consumption by private households. Despite the growing number of appliances that consume power in standby modes, the studies indicate that standby power consumption in Germany could be reduced to 8 TWh per year or less within ten years if adequate measures are taken.

STANDBY POWER MODES

In the last decade, power-saving standby modes have been introduced, particularly for products that are permanently plugged in. These devices reduce the total energy consumption of the appliances. Nearly all appliances manufactured today have one or more standby modes in which the product draws less power when it is performing only certain functions or when it is waiting to provide full service. Unfortunately, for some products, such as computers, power management features that are incorporated in the product design are switched off before shipping because of concerns over functionality, which are often unfounded.

Most electrical products, such as televisions, VCRs, audio products, cordless telephones, computers and power speakers, have a power switch that allows the user to turn off the product or put it into a standby mode, sometimes called the "idle mode". Parts of these devices remain in the standby mode until they receive input from a remote control device or activation of the power switch.

There are problems with networked appliances that make power management by the user difficult. Some industrial consumers disable their personal computer (PC) standby modes because the network manager wants permanent access or because of communication problems between PCs and printers. These problems need to be addressed to avoid having appliances remain "on" 24 hours a day despite product labels indicating standby mode savings.

Some products require the user to perform two actions to bring a product up to its full operating mode. One is to initiate a mode from which the user may choose from a number of services. In most cases, the device returns to this state when it has completed the requested service. Examples of products of this type include tape and compact-disk (CD) players, VCRs, printers and some dishwashers.

Some standby modes, commonly referred to as "sleep modes", improve energy efficiency in products that consumers frequently leave on when they are not using them. Such products can be programmed to turn off selected components after a period of non-use. Some of the earliest products to use standby modes were portable laptop computers, which automatically turn off key components when the user does not use the keyboard or mouse for a predefined period of time. Today, most computers have two power-saving modes. The first kicks in when the active computer is not used for a certain prescribed period of time, often 30 minutes. The computer enters a second, "deeper" sleep mode if it remains inactive beyond the initial time period, say 45 minutes since the last user input. Sleep modes could be introduced in many other products as well, such as stereos, digital-versatile-disk (DVD) players and digital television decoders – also called set-top boxes. In computers and other office equipment, power-saving standby modes have been introduced by manufacturers as a result of voluntary government programmes to save energy. In the future, power-saving standby modes and power management should become standard features of every networked appliance.

For some devices, consumers do not have the option to initiate standby modes, either because mode changes are automatic or because the manufacturer decided that a standby mode was inappropriate. For example, exit signs and electric clocks are expected to be "on" or actively functioning all the time. Some argue that standby modes for these devices would be impractical, while others believe that automatic activation, by motion or voice, is a possibility. Similarly, devices such as doorbells, remote switches, smoke alarms, carbon monoxide detectors and thermostats are in a "ready" state most of the time. Whether this ready state should be categorised as a standby or active mode is controversial. Newer products, such as certain set-top boxes and network devices, need to be ready for information processing at all times and so are not designed to be switched off. The appropriateness of standby modes must be carefully considered on a product-by-product basis.

Some products consume power even when they are not doing anything. For instance, electricity from the mains continues to flow into battery chargers and some audio products as long as they are connected to a socket. On the other hand, televisions without remote control capabilities (now quite rare) and most power amplifiers do not perform any standby functions and do not draw power when they are off. In some devices the standby mode has about the same power consumption as the active mode and so switching to standby modes does not save energy.

Characteristics of standby modes differ from product to product and from country to country. Some countries have special appliances that have standby power consumption. For example, in Japan, "shower-toilets" consume 5 watts and rice cookers consume 1 to 2 watts. In France, Minitel communication systems draw 5 to 9 watts. However, the large majority of products with standby consumption are internationally traded goods.

TECHNICAL SOLUTIONS

INTRODUCTION

A significant number of electronic appliances spend most of their lives in standby mode. In some cases, standby energy use is several times the active energy use over the lifetime of the appliance. The most notable example of this is the VCR, which on average consumes 19 times more electricity in standby mode than while actively recording or playing[1]. For such appliances, reducing the power requirements of standby modes even slightly may achieve significant energy savings. An alternative strategy is to minimise the amount of time that appliances are in the high-power mode by ensuring that the appliance is always in the lowest possible power mode for the desired level of operation. This principle is known as power management.

BARRIERS TO STANDBY EFFICIENCY IMPROVEMENTS

It is often difficult to estimate the costs of improving energy efficiency. This is especially true when the improvement is part of an already-planned redesign cycle to change the way the product functions or to reduce costs. At issue may be design costs, component costs or manufacturing costs. There are still other cost factors that are beyond the control of the manufacturer.

1. Rosen and Meier, 2000.

In some situations, product prices are not set by the manufacturer. For example, set-top boxes, or integrated receiver decoders (IRDs), are provided to the consumer by a service provider as part of a service package. The service provider sets the price it is willing to pay for the IRD. Thus, the manufacturer is asked to make the set-top box as cheaply as possible within the design specifications of the service provider.

Another barrier is price sensitivity in the highly competitive electronics market-place. In most retail stores, a price increase of even a few euros or dollars cannot be passed on to the cost-conscious consumer. Retailers are reluctant to increase the price of a product from, for example, $199 to above $200. A small price increase may push many products out of the consumer's perceived price range relative to other available products, and cost the retailer a sale. Theoretically, a price rise would be offset by the benefits to the consumer of reduced energy bills from using energy-efficient appliances. But most consumers are not aware of these benefits or are more concerned with the upfront cost of an appliance.

Even when efficiency improvements have minimal impact on pricing, many manufacturers will ignore product efficiency for other reasons. For example, with limited time and workforce, priority may be given to factors such as improved functionality rather than energy efficiency.

Maintenance costs can also constitute a barrier to efficiency improvement. The more efficient systems and appliances become, the more complex and therefore expensive they can be to maintain.

POWER SWITCH PLACEMENT

One simple technical solution to unnecessary standby power consumption involves the placement of the switch in relation to the power supply. The type and placement of the switch dictate whether a device that supplies power to an appliance has to consume energy when the appliance is not being used. Figure 3.1 shows some common ways to design power switches in electrical products. The power supply can be internal (in the appliance) or external (outside the appliance).

Figure 3.1 **Common power switch designs**

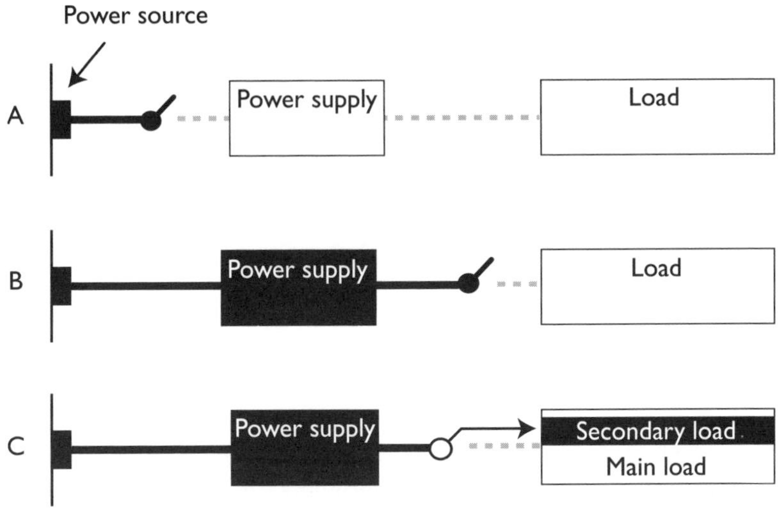

Source: Lawrence Berkeley National Laboratory.

When a mechanical switch is placed between the power source and the power supply (Figure 3.1, diagram A), no power is consumed when the appliance is switched off because the current does not reach any energy-consuming components.

When a mechanical switch is placed between the power supply and the appliance (on the low-voltage side of the power supply), some energy will be consumed even when the appliance is switched off (Figure 3.1, diagram B). Some of the current that flows into a power supply is dissipated as heat, even when the appliance is turned off or when there is no appliance connected to it. Appliances with switches that are designed in this way, such as some audio products and halogen desk lamps, consume energy as long as they are connected to the power source, even though they are not fulfilling any function at all. This power use can be avoided by repositioning the switch between the power source and the power supply as in diagram A. There could be other reasons, such as safety or costs, for not locating the power switch as shown.

Sometimes it is not feasible to completely eliminate standby power consumption – when, for instance, a secondary load (such as a clock) needs to be energised (diagram C in Figure 3.1). In such cases, three major options exist for minimising standby power use: adding an extra power supply for use at low-power levels, using a power supply with two operating ranges (one for the "on" mode and one for the standby mode) or incorporating a separate power source such as a small battery or photovoltaic cell.

IMPROVING COMPONENT EFFICIENCY

Two other approaches to reducing standby power consumption are to improve the efficiency of components that function while in standby mode, or to use new or different components that require less power. Such components include low-voltage power supplies with voltage regulators, integrated circuits and displays. Options for improving the efficiency of the components that use significant amounts of standby power are discussed below.

Power supplies

Some types of power supplies are much more efficient than others. The two main types are linear power supplies and switch-mode power supplies. Some linear power supplies are fairly efficient, but switch-mode power supplies tend to be even more efficient. The choice of power supply can have a major impact on standby power consumption.

It may be possible to reduce standby power consumption by adding a separate, secondary power supply that consumes less energy than the power supply that is used by the appliance in its active mode. An alternative may be to incorporate a small battery or photovoltaic cell.

Voltage regulators

In most electronic products, voltage regulators dissipate as heat a large percentage of the power that flows through them. Sometimes this dissipation accounts for a higher degree of power consumption than that of any other single component. Dissipation of power increases the temperature and so shortens the lifetime of appliances. As with power supplies, some types of voltage regulators are much more efficient than others. It may be possible to reduce overall standby power by using a more efficient voltage regulator. Another option is to use fewer different voltage levels, so that fewer voltage regulators are needed. Low dropout (LDO) voltage regulators are more efficient than standard regulators because they dissipate a much smaller fraction of the power.

Integrated circuits

Another option for reducing standby consumption is the use of efficient integrated circuits. Efficient integrated circuit components have been designed for battery-powered products. These might be adaptable for use in products that are powered from mains. Higher system efficiency may also be achieved by switching to a lower-frequency oscillator while the appliance is on standby.

Visual displays

Power savings could be realised by changing the type or size of visual displays. The most common luminescent display panels in appliances are light-emitting diodes (LEDs) and vacuum fluorescent displays. These displays come in the form of round or oval lights that can be used alone or assembled to create shapes and characters. A commonly used non-luminescent display is the liquid crystal display (LCD), which is typically black and white. In some appliances, such as portable computers, LCDs are enhanced with colour and background lighting. In general, black-and-white LCDs are the most efficient of these three types of display and could reduce standby power consumption, although they may not be appropriate for all applications.

Where the use of efficient LCDs is not practical or cost-effective, it may still be possible to save some power by using smaller displays. Lower-power LED technologies are also becoming available.

WILL REDUCING STANDBY POWER RAISE THE PRICE OF PRODUCTS?

Any change in the design and construction of a product may modify its production cost and, eventually, the price consumers pay. The costs to reduce the standby power consumption of most appliances are surprisingly modest and, in many cases, may result in lower costs or new benefits elsewhere.

The main benefit of reduced electricity use is simply expressed as the avoided electricity bills for the life of the device. The present value of saving one watt of standby power ranges from $3 to $8, depending on the local electricity price. This means that a consumer should not pay more than about $5 for each watt saved. For a 5-watt saving, the incremental cost should be less than about $25. The present value of savings establishes a way of quickly identifying cost-effective strategies to reduce standby power.

One must also consider the mark-up that occurs from manufacturer's cost to retail price. The mark-up can be anywhere from 300 to 500 per cent. Thus, an increase of $1 in the manufacturer's cost will lead to as much as a $5 increase in the retail price. The mark-up is an especially important consideration for relatively inexpensive appliances such as desktop halogen lights. These units often consume standby power because the manufacturers have inserted switches on the low-voltage side of the power supply. The no-standby alternative, putting a high-voltage switch on the high-voltage side, adds about one dollar in manufacturing costs. Many manufacturers use the low-voltage switch because they know price-conscious consumers would buy another model rather than pay an extra $3.

There are three strategies to reduce standby power: improve the efficiency of the power supply and other components, use lower-power components and de-energise components not required during standby mode. Each has unique costs and benefits, a few of which are described below.

Manufacturers can improve the efficiency of the power supply by converting to switch-mode power supplies. The extra cost of switch-mode units (beyond the cost of traditional designs) ranges from nothing to about $1, depending on the size and other requirements. The price differential is falling rapidly. The simple drop-in replacement strategy will typically save 1 to 3 watts, so they are already cost-effective. Switch-mode power supplies also operate more efficiently in the "on" mode; this may result in additional savings. Depending on the situation, the payback time (for consumers) will be very short, often only a few months and almost always less than a few years.

Inserting a switch-mode power supply may lower other costs. For example, one company found that it could lower inventory costs because it could use the same power supply in all countries. Another firm found that the lighter switch-mode power supplies made the product more durable and less likely to separate from the chassis when dropped. This resulted in fewer call-backs and repairs. Many firms ship their products by air; since switch-mode power supplies weigh less than traditional designs, the firms save on freight costs.

A second strategy involves using lower-power components. The most common example occurs in displays. New, low-power displays using LED technologies are now becoming available in a range of colours and intensities. Complex displays that use little standby power will cost more, but LED technologies are already cost-effective for many applications. And the prices are falling rapidly.

A third strategy involves de-energising components that are needed during the standby mode. This strategy is often called power management. Many home electronics rely on sophisticated chips to control their operation and have power management features already built into them. Manufacturers have often failed to exploit those features because they were in a hurry to get products to market. The incremental cost of enabling power management features is often very low; sometimes it is just a matter of asking the engineers to do it. For set-top boxes and other devices relying on a service provider, the increased cost is mostly to achieve communications protocols that will permit a device to switch off components until they receive a signal to wake up.

CURRENT POLICIES TO ADDRESS STANDBY POWER

INTRODUCTION

In adopting policies to encourage incorporation of energy-efficient features in product design, policy-makers have focused on standby power consumption as an area where real progress could be achieved in a relatively short period of time.

Over the past decade, governments have tried to reduce standby power in various ways. Many IEA countries already have programmes to reduce standby power in the most used appliances, such as televisions, computers, VCRs and audio equipment. Regulatory standards, labelling and voluntary agreements are the main policy approaches that have been used to address standby power to date.

DEMAND-SIDE MANAGEMENT PROGRAMMES

UK Market Transformation Programme[2]

The UK Market Transformation Programme (UKMTP) is a government programme run by the Department of the Environment, Transport and the Regions in the United Kingdom. This programme focuses on providing information on energy consumption to manufacturers, retailers and consumers and

2. *www.mtprog.com*

helping them to make purchase decisions. The UKMTP aims to develop a consensus among stakeholders on market projections for energy consumption. The consensus-building process focuses on both active and standby energy consumption and involves manufacturers and retailers, utilities, government departments and agencies, the Energy Savings Trust and consumers. Scenarios are developed for 10- to 20-year periods. These are used to design appropriate measures to limit energy consumption. Sectoral review papers provide information on equipment stocks, equipment usage, technological developments, and projections of energy use and carbon emissions. Discussion of these papers helps develop consensus.

The UKMTP covers 27 product groups, of which over half have standby modes. The programme addresses the power consumption in both standby and active modes of most domestic energy-consuming appliances, including televisions, refrigerators, ovens and lighting, and of traded goods in the commercial sector, including office equipment, electric motors and lighting.

IEA Demand-side Management Implementing Agreement, Annexes III and VII[3]

In many IEA countries there has been growing interest in the concept of "market transformation" and a growing number of promising new activities. Market transformation is a term used in the IEA Implementing Agreement[4] on Demand-Side Management (DSM). It refers to public policies and programmes of limited duration that promote the availability, marketing, sale and use of energy-saving equipment or practices by creating a permanent change in market structure or processes.

3. dsm.iea.org

4. "Implementing Agreements" are collaborative activities in technology R&D, technology analysis, and information sharing and dissemination that are carried out under the aegis of the IEA. "Annexes" are the specific tasks carried out under an agreement.

Many of these market transformation activities focus on creating demand for energy-efficient products as well as an adequate supply of them. Aggregated procurement strategies and energy labelling are two of the important themes in these new programmes. A third is the creation of the infrastructure (design, installation, financing and maintenance) to support new energy-saving technologies in the market-place.

Increased involvement of retailers in the distribution and sale of more efficient products is important. Retailers can exercise early influences on the formulation of specifications for new products. They can also direct advertising and sales efforts for the more efficient products. The market transformation programmes' success depends in a large part on the retailers' willingness to promote more efficient products.

Since 1993, several pilot procurement projects have been carried out under Annex III of the DSM Implementing Agreement. At least two projects are likely to continue, including one on next-generation office copiers. The IEA is now embarking on a new initiative on International Collaboration on Market Transformation (Annex VII). This initiative will seek to increase the market share of energy-saving products and accelerate the use of the most efficient new technologies.

In order to build on the experience gained from Annex III and other activities under the DSM Agreement, the new initiative focuses on energy rating, labelling, quality marks, and procurement of energy-efficient products. The work also comprises study of market needs on both the supply and demand sides. Annex VII of the agreement will also stimulate complementary activities that reinforce demand, competitive supply, and the proper installation and use of energy-efficient products. The initiative will include information exchange and research, as well as a series of co-operative market transformation projects. The flexible approach will allow countries to pursue a wide range of market transformation projects together with other interested countries. Standby power consumption will be included in the design of activities to promote these products.

REGULATION

Japan's Top Runner programme

The only existing standard for standby power that is purely regulatory is the Japanese Top Runner programme, which was established in March 1999 under Japan's framework legislation on energy efficiency.[5] The Top Runner programme sets energy efficiency targets for 11 products. It covers passenger vehicles as well as air-conditioners, heaters, fluorescent lamps, television receivers, copying machines, computers, magnetic disk devices, VCRs, refrigerators and freezers. Several other products that consume a large amount of energy are being considered for inclusion in the programme.

Top Runner standards will become mandatory at specified future dates. For example, the energy efficiency of computers must be improved by 83 per cent by 2005. The programme divides products into sub-groups, taking into account factors such as size, weight and function. The performance of the most efficient appliance on the market is set as the minimum standard to be met at a specified future date. This standard will apply to both domestic and imported products. Expected changes in technologies and environmental regulations are taken into account in setting the standards. Targets will be made even more stringent than the highest current level of energy efficiency if large improvements are considered likely by the target year. For many products, such as televisions, copiers and computers, the Top Runner standard is based on the full-service cycle, including energy consumption in standby and "on" modes. (VCR standards are based on the standby mode only.)

If a manufacturer cannot reach the target by the target year, the Ministry of Economy, Trade and Industry (METI) will issue

5. The "Law Concerning Rational Use of Energy" (Energy Conservation Law) enacted in 1979.

recommendations. If the manufacturer fails to abide by the recommendations, the name of the firm will be made public or an administrative order will be issued. So there is a very strong incentive for manufacturers and importers to meet the Top Runner targets. Compliance is assessed by comparing actual and target performance across a weighted average of products made by the manufacturer, rather than product by product.[6]

The standards are challenging and were only agreed to after intensive discussions and negotiations between the government and representatives of Japanese industry sectors. These discussions took place in advisory bodies set up by METI. This consultation process was difficult, but growing awareness of global warming since the Kyoto climate change conference in 1997 helped the participants to reach a conclusion within one year. The standards were then promulgated under existing legislation.

An energy efficiency labelling scheme for household electric appliances is also being launched along with the Top Runner programme. The aim is to support manufacturers who meet the targets and to disseminate information on the energy efficiency of different products to consumers. The labelling scheme covers air-conditioners, fluorescent lamps, television receivers, refrigerators and freezers. The new labels will indicate the degree to which the Top Runner target has been achieved, in percentage terms. The objective is to influence consumer choice and accelerate manufacturers' efforts to achieve the target well ahead of the target year.

The Top Runner programme defines its criteria in a different way from programmes in other countries and so cannot be directly compared with other standards, voluntary agreements and labelling schemes.

6. Jun Arima, METI (paper presented at UNFCCC Best Practice Workshop, Copenhagen 11-13 April 2000).

Swiss regulations

Under Swiss energy legislation, voluntary agreements are the first phase of regulation, and ordinances are put in place to enforce energy efficiency standards if voluntary agreements do not meet their objectives. Two trade associations have entered into voluntary agreements with the Swiss government that relate to standby power consumption. These associations are made up of enterprises in the business and consumer electronics markets, the computer and software market and the household appliance market (FEA). Twelve ordinances governing standby and "off" mode energy consumption were put in place between 1993 and 1995 with target dates from 1995 to 1999.

In many cases these targets were not reached. Targets were met by 53 per cent of household appliances and 87 per cent of office equipment. No printers reached the standby power consumption target of 2 watts, although 40 per cent fell below 4 watts. Average standby power consumption for new printers in Switzerland fell from 17 watts in 1994 to 7 watts in 1997.

VOLUNTARY INDUSTRY STANDARDS

As more products are traded across borders, the use of voluntary industry standards is becoming a prerequisite to global trade. Industry benefits from the predictability and uniformity that standards create. The current trend is towards international standardisation, particularly through the ISO (International Organization for Standardization). Electrotechnical standards are harmonised internationally by the ISO and by the International Electrotechnical Commission (IEC). Both government and industry experts participate. Trade associations also ensure that the views of their members are represented.

Committees of manufacturers, users, research organisations, government departments and consumers work together to draw up standards that evolve to meet the demands of society and technology. In many cases, standards processes are organised by industry trade associations and other industry bodies rather than by governments.

There are many examples of voluntary industry standards, where companies develop and adopt test methods and standards for products. Voluntary standards can be a very effective alternative to regulations and other government-imposed measures for products that are traded widely. Industry standards organisations seek to ensure that the voluntary industry standards they produce are in line with consumer requirements and are acceptable to regulators.

One example of a voluntary industry standards body is ASHRAE, the American Society of Heating, Refrigerating and Air-Conditioning Engineers. ASHRAE writes standards and guidelines for industries throughout the world in the delivery of goods and services. ASHRAE produces uniform methods of testing, recommended practices for designing and installing equipment, and other information. ASHRAE has more than 87 active committees on standards and guideline projects that address a very wide range of cases, including indoor air quality, thermal comfort, energy conservation in buildings, refrigerant emissions, and the designation and safety classification of refrigerants.[7]

Another example is the American Association of Mechanical Engineers (ASME). Since 1884, ASME International has pioneered the development of codes, standards and conformity-assessment programmes. ASME maintains and distributes 600 codes and standards used around the world for the design, manufacture and installation of mechanical devices.[8]

---

7. www.ashrae.org
8. www.asme.org

The Association of Home Appliance Manufacturers (AHAM) is a trade organisation of home appliance manufacturing companies. AHAM provides industry statistics, product certification and industry self-regulation. AHAM has developed standards for many household products including dishwashers, washing machines and humidifiers.[9]

The British Standards Institution (BSI) is a non-profit organisation and is globally recognised as an independent and impartial body serving both the private and public sectors. BSI does not write standards but brings qualified parties together who are interested in developing a particular standard. Any manufacturer can claim that its product meets the relevant BSI standard but it can be prosecuted if the claim proves to be false.[10]

In Japan, industrial standardisation is promoted at the national, industry association and company levels. JIS (Japanese Industrial Standards) are government-endorsed but voluntary national standards for industrial and mineral products. Various industry associations also establish voluntary standards for their specific needs. The JIS operates a voluntary certification system. Factories manufacturing products that satisfy JIS are allowed to use the JIS mark on their products. The aims of JIS and the JIS marking system are to improve the quality of products, rationalise production, and ensure fair and simplified trade. Mandatory regulations, such as technical regulations for electronic appliances, are promulgated under the Electrical Appliance and Material Control Law, but these are always in line with JIS.[11]

---

9. www.aham.org
10. www.bsieducation.org
11. www.hike.te.chiba-u.ac.jp/ikeda/JIS

CONSUMER INFORMATION: VOLUNTARY LABELLING

Energy efficiency labelling programmes

Energy Star

Under the US Energy Star programme, the government, manufacturers and other key players such as energy utilities and retailers agree to meet specified energy efficiency criteria. Since the Energy Star programme began in 1992, the US government has forged partnerships with over 1 200 firms. Manufacturers who sign an Energy Star agreement must develop at least one product that meets Energy Star criteria and can place the Energy Star label on products that meet the criteria. The programme also educates consumers about the benefits of energy-efficient products.

The Energy Star programme includes energy consumption criteria for 31 different types of products including a wide range of office equipment, consumer electronics, and home heating and cooling equipment.[12] The US Environmental Protection Agency (EPA) is considering expanding the programme to other products, such as set-top boxes, ventilation and ceiling fans, telephones, water coolers and motors.

There is a procurement component to the Energy Star programme. The federal government requires its departments to buy Energy Star office equipment. Private companies have also signed memoranda of understanding with the EPA to purchase Energy Star office equipment and place Energy Star specifications in their contracts with suppliers. The enormous demand pool that this requirement creates influences the design of personal computers and other types of office equipment that are sold not only in the

12. *Energy Star web site: www.epa.gov/appdstar/estar/ prod_dev.html*

United States but worldwide. In 1997, 82 per cent of computers, 92 per cent of monitors, almost all printers and 95 per cent of fax machines that were sold in Switzerland complied with Energy Star criteria.[13]

Japan, New Zealand, Taiwan and Australia have adopted the Energy Star labels and criteria for their energy efficiency programmes. Canada, Brazil, Mexico and other countries have also expressed interest in the programme. In Australia, the focus of the Energy Star programme has been on office equipment, though government and industry have recently agreed to expand the programme to include home entertainment equipment.

The US EPA and the European Commission are completing negotiations on a co-ordinated Energy Efficient Labelling Programme for Office Equipment. This agreement is expected to enter into force after the European Parliament has adopted a regulation regarding the implementation of the office equipment labelling programme in Europe. Under the agreement, new specifications can be put in place only if they are accepted by both the United States and the European Union. A common set of energy efficiency specifications and a common logo (the Energy Star logo) will be used by the United States and the European Union. A key objective of the agreement is to maximise the effect of individual countries' efforts to promote the supply of, and demand for, efficient products.

The Energy Star logo is recognised by consumers in many countries, and the label is easy to understand. According to a survey by the American Council for an Energy Efficient Economy (ACEEE), 80 per cent of consumers who are concerned about energy efficiency said they were familiar with the logo, while 43 per cent said they had used the logo in purchasing a product. The cost of

13. Zielwerte von Elektrogeräten: Auswertung der Datenerhebung 1997, *Eric Bush, Bush Energie, Felsberg, Switzerland, 1998.*

achieving Energy Star efficiency levels is estimated by manufacturers to be negligible for many appliances.

Group for Efficient Appliances label[14]

The Group for Efficient Appliances (GEA) is a forum of representatives from national energy agencies and governments who are working with industry on voluntary information activities to improve the efficiency of electronic products, mainly household appliances. The GEA label is part of a voluntary programme that was started in 1996. Energy agencies from eight European countries (Austria, Denmark, Finland, France, Germany, the Netherlands, Sweden and Switzerland) are members of the GEA. Other agencies or organisations can participate under specific conditions. Information is sent out to all interested parties, such as manufacturers and importers, and the European Commission.

Each GEA member carries out information campaigns appropriate to its consumer market. The participants exchange information on current and planned activities. Test methods are harmonised as much as possible with other labelling schemes, such as Energy Star. The GEA criteria are revised regularly in co-operation with industry.

The GEA programme covers televisions, VCRs, television-VCR combinations, IRDs (set-top boxes), DVD players, audio sets and audio components, wall packs and battery chargers, personal computers, monitors, printers, mailing machines, fax machines, scanners and copiers. So far, this programme has addressed standby power consumption but not "on"-mode consumption. It will include "on"-mode power consumption for televisions from January 2002. GEA labels are awarded only to appliances that are within the top 20 to 30 per cent of all models available in the combined market of those countries taking part in the GEA

14. *www.gealabel.org*

voluntary agreement. The criteria for receiving the GEA label were established in co-operation with industry.

An Internet database provides information on all products that have earned the label. Each country carries out activities to promote those products, by maintaining a national web-site or distributing lists.

Multi-attribute labelling programmes

EU eco-label

The voluntary European Union eco-label defines criteria on the basis of a "cradle-to-grave" assessment of the environmental impact of the product group. The eco-label addresses both "on"-mode power consumption for white goods and light bulbs, and standby power consumption for computers. Eco-label criteria are defined for personal computers and monitors, portable computers, washing machines, dishwashers, tumble dryers and refrigerators. The European Commission develops ecological criteria for product groups in close collaboration with EU member states. The Commission cannot adopt criteria before environmental experts have given their opinion through the Eco-label Regulatory Committee. Key interest groups must be consulted about proposals for the definition of product groups.

Blauer Engel

The German Blauer Engel (Blue Angel) eco-label is part of a voluntary programme that provides standards for a range of environmental impacts. Currently, the scheme covers personal computers and monitors, portable computers, printers, fax machines, copiers, televisions, washing machines, dishwashers, tumble dryers and refrigerators. The programme is carried out under the German Ministry for the Environment. The labelling criteria specify maximum energy consumption for each product group in different modes of operation. It also sets criteria such as the maximum "wake-up time" from the energy-saving mode and the maximum time required for energy-saving features to switch on. Product groups are categorised by function. Manufacturers who are certified as meeting the criteria can apply the label to their products.

Nordic Swan

The Nordic Swan scheme is another voluntary eco-labelling programme covering a range of environmental impacts. It is organised by the Nordic Council of Ministers and is used in Finland, Norway and Sweden. The product types that are covered by the programme are personal computers and monitors, portable computers, printers, fax machines, copiers, televisions, VCRs and television-VCR combinations, stereo systems, washing machines, dishwashers and refrigerators. The criteria for the standby power consumption of office equipment are equivalent to those of Energy Star or GEA. For several of the consumer electronic appliances the "on" mode is taken into account as well as standby modes. The Nordic Swan eco-label is administered by national boards that co-operate through the Nordic Ecolabelling Board.

Activities to reduce standby power in other countries

Germany

In Germany over the past few years, standby power consumption has become a subject of concern for both the federal government and local authorities. Berlin has issued several resolutions concerning the reduction of standby power losses. The government supports a 1-watt goal for standby power and recommends the principle, "off is off".

Some utility companies, especially the ones belonging to ASEW (Association of Municipal Public Utilities for Energy Efficiency), run regular communication campaigns to convince consumers to pay attention to the problem of standby power consumption. This includes an award of DM 50 for the purchase of appliances with low standby power losses. The GED (Gemeinschaft Energielabel Deutschland – German Energy Label Association) has endorsed the GEA label and promotes this scheme throughout the country.

Switzerland

Energy 2000[15] is the Swiss national counterpart to the GEA labelling scheme. The label covers office equipment (personal computers, monitors, printers, copiers, scanners, fax machines, multi-functional devices and energy-saving devices), consumer electronics (televisions, VCRs, hi-fis and DVD players) as well as water heating equipment, lighting products, multi-socket adapters and battery chargers.

Energy 2000 awards labels to the top 25 per cent of units on the market in a given year in terms of energy efficiency. The Zurich city council and at least one major bank in Zurich now purchase only Energy 2000 equipment. According to the Swiss federal government managers of the programme, no price increases have been observed as a result of the labelling scheme. The cost has been absorbed by wholesalers and retailers within their profit margins.

France

In September 1998, France's secretary of state for industry asked the national standards body (AFNOR, Association française de normalisation) to develop a series of standards aimed at reducing energy consumption (including energy in standby modes) in home and office electronics. Activities started in 1999 under the aegis of AFNOR and the National Energy and Environment Agency (Agence de l'environnement et de la maîtrise de l'énergie – ADEME) and the first results are expected in 2001.

15. *www.energielabel.ch*

Australia

After consulting with industry, the Australian government has developed a strategy to reduce standby power consumption. The basic approach is to support international co-operative programmes, and to undertake programmes to help reduce standby power in domestic and imported appliances. In April 2000, the Council of Commonwealth, State and Territory Ministers in charge of energy matters (ANZMEC) endorsed a programme of work designed to lead Australia towards achievement of a 1-watt standby power target for all products. Consequently, policies are designed that ensure that the maximum standby power consumption of all appliances manufactured in or imported into Australia is one watt. This statement of principle sends a clear message to industry and provides coherence to a diverse range of policies designed to combat standby power consumption.

Until recently, Australia's focus was on office equipment. Existing policies include an Energy Star programme for office equipment and the provision of information to businesses to help them buy and operate energy-efficient office equipment. However, the emphasis has now moved to domestic appliances, with an agreement to expand the Energy Star programme to include home entertainment equipment. There is also a commitment to incorporate standby power consumption into the existing Energy Rating Labels for white goods.[16]

Australia is committed to the international harmonisation of standby test procedures. It has demonstrated its support by funding the chair of the IEC TC59 ad hoc working group, which is currently developing a test method to measure standby power consumption.

---

16. www.energyrating.gov.au

VOLUNTARY AGREEMENTS

European Association of Consumer Electronics Manufacturers (EACEM)[17]

EACEM and the European Commission have concluded voluntary agreements to reduce the standby energy consumption of colour televisions and VCRs. Manufacturers who have signed the agreements hold more than 80 per cent of total market share for these products in Europe. Under these agreements, from January 2000 the maximum standby power consumption of these products is 10 watts. The agreements specify that the sales-weighted average standby power consumption of all units sold by a given manufacturer may not exceed 6 watts. The standby power levels of televisions were already well below the target of 6 watts by 2000, an indication that the targets were already incorporated in product design. Market figures show a continuously declining level of standby power consumption in European televisions, from 7.5 watts in 1995 to 3.7 watts in 1999. It is likely that the voluntary agreement process has contributed to this trend.

European codes of conduct

The European Commission and two trade associations, EACEM and EICTA (European Information and Communication Technology Industry Association), have finalised in year 2000 a voluntary code of conduct that will reduce the energy consumption of audio systems. The associations have recommended that their members sign the code. This agreement covers standby power consumption. Targets negotiated under this agreement are 5 watts by January 2001, 3 watts by January 2004 and 1 watt by January 2007.

17. EACEM members are: Aiwa (UK) Ltd. (indirect member via BREMA), Bang & Olufsen, Grundig AG, Hitachi Home Electronics Europe Ltd, JVC Europe, Loewe Opta GmbH, Mitsubishi Electric UK Ltd, Panasonic Europe Ltd, Philips Sound & Vision, Pioneer Electronic Europe NV, Samsung Europe Headquarters, Sanyo UK Sales Ltd, Sharp Corporation, Sony Europe GmbH, Thomson Multimedia, and Toshiba.

The European Commission is also negotiating codes of conduct for integrated receiver decoders (IRDs) and external power supply units (PSUs) with key stakeholders. These codes of conduct focus on standby power consumption. The EU maximum power levels for IRDs are 6 watts in "standby-passive" mode and 9 watts in "standby-active, low" mode.[18] The European Union is seeking to introduce these codes of conduct by 1 January 2003. For wall packs and battery chargers, the agreement asks signatories to attain a no-load power consumption of less than one watt by 1 January 2001 and less than 0.75 watt by 1 January 2003.

By June 2000, after more than two years of discussion, only a few companies had signed the code of conduct for IRDs. By December 2000, industry had begun to sign the code of conduct for PSUs. For IRDs, several different parties – service providers, IRD manufacturers, silicon manufacturers – must co-operate to achieve the targets, and the technology needed to achieve the targets is not as simple as it is with televisions or VCRs. These factors make it difficult to agree on targets. Moreover, if proposals from manufacturers are too weak, no significant savings will be achieved. But if targets are too strict, voluntary programmes will not attract participants.

THE EFFECTIVENESS OF CURRENT APPROACHES

Voluntary labelling programmes are the most widely used policy approach for addressing standby power, both in geographical coverage and in the number of appliances covered. The Top Runner programme (Japan) is the only mandatory programme that addresses standby power consumption specifically.[19]

18. The difference between standby-passive and standby-active, low is that in the standby-active, low mode the IRD can be switched into another mode by an external signal, such as a service provider who wants to download software during the night.

19. Some existing programmes focus on total energy use, which includes standby energy although standby energy is not the main element.

Despite the success of various labelling programmes, standby power consumption can still be substantially reduced. Current policies do not cover all the products that consume power in standby modes, notably telephones and telephone chargers, including the ubiquitous mobile phone chargers. Moreover, standby consumption is expected to increase with the development of networked home and office products. Government actions will need to be broad and flexible to keep up with fast developments in product markets, including the dramatic growth of electronic commerce.

A positive development in the last decade has been the growth of co-operative processes and partnerships between government and industry. Processes involving government, product and component manufacturers, service providers, retailers and consumers are increasingly common in policy deliberations. Policy tools falling short of regulation are being used to achieve the same goals without compulsion. While regulatory approaches have achieved some success, they tend to be controversial and slow to adapt to rapidly changing product markets.

The programmes that have been implemented to date are not globally consistent. Most do not specifically address standby power use. Some schemes seek to motivate manufacturers to produce efficient products by establishing minimum standards. Others seek the same goal by withholding recognition of all products except those with the highest standby efficiency. Definitions, methods, criteria and test procedures often vary from one programme to another, forcing manufacturers to design and develop multiple versions of the same products in order to comply with the various requirements.

Differing national schemes can greatly increase the cost of manufacturing and selling products in the global market, while producing little benefit for the environment. A single harmonised labelling programme would simplify compliance with environmental criteria while reducing administrative and production costs. This would create a more constructive and co-operative relationship between government and industry. A globally harmonised label may, however, be possible for only a small number of appliances.

The electronics market is moving very fast. Efforts to reduce standby power consumption need to keep pace with market developments. If governments and other actors fail to take market developments into account, efforts to reduce standby power will lag behind the expected proliferation of equipment that consumes too much power in standby modes.

Table 4.1 compares a number of different national and regional schemes. Table 4.2 illustrates the product coverage of different policy initiatives.

Table 4.1 Comparison of different programmes

Product group	Standby energy criteria (maximum power consumption in watts)					
	Swiss ordinances	Energy Star	GEA*	EU eco-label	Blauer Engel	Nordic Swan
Monitor	3 W	"sleep" mode: 15 W "deep sleep" mode: 8 W	3 W (without USB), 5 W (with USB)	3 W	5 W ("off" mode: 1 W)	8 W** ("on" mode: depends on screen size)
Personal computer (system unit)	13 W with integrated monitor 10 W without integrated monitor	30 W	equal to Energy Star ("off" mode: 5 W)	27 W (stand alone)	30 W* ("off" mode: 1 W)	30 W** ("on" mode: 230 W)
Portable computer				5 W ("off" mode: 3 W)	7 W ("off" mode: 2 W)	8 W
Printer (laser/LED)	2 W ("off" mode: 1W)	0 to 7 pages per minute (ppm): 15 W 7 to 14 ppm: 30 W > 14 ppm and high-end colour: 45 W	equal to Energy Star (large size, colour)		equal to Energy Star ("off" mode: 2 W)	equal to Energy Star
Printer (ink jet, matrix)	2 W ("off" mode: 1 W)	0 to 7 ppm: 15 W 7 to 14 ppm: 30 W > 14 ppm and high-end colour: 45 W	for standard size: ≤7 ppm: 6 W >7 ppm: 16 W			equal to GEA
Fax machine	2 W	0 to 7 ppm: 15 W 7 to 14 ppm: 30 W > 14 ppm and high-end colour: 45 W	2 W		<7 ppm: 7 W ≥7 ppm: 15 W	equal to GEA
Copier	27 W ("off" mode: 1 W)	low power mode: 0 W to 175 W (depending on ppm) "off" mode: 5 W to 20 W (depending on ppm)	equal to Energy Star		equal to Energy Star ("off" mode: 2 W, 5W)	equal to Energy Star

Product group	Standby energy criteria (maximum power consumption in watts)					
	Swiss ordinances	Energy Star	GEA*	EU eco-label	Blauer Engel	Nordic Swan
Television	3 W	3 W	1 W (from 1-1-2002, standby included in duty cycle)		4 W	1 W*** (100 Hz TV: 3 W) ("on" mode: depends on screen size)
Television-VCR combination		6 W	3 W			4 W ("on" mode: depends on screen size)
Stereo system		Home audio: 2 W when "off" DVD: 3 W when "off" (1 W for home audio and DVD products from 2003)	2 W			3 W*** ("on" mode: 40 W)
VCR	3W	4 W	3 W			2 W ("on" mode: 15 W)
Washing machine				EU energy label A	EU energy label A end of cycle: 5 W "off" mode: 1 W	"on" mode: average of various cycles
Dishwasher				EU energy label A, B or C	EU energy label A end of cycle: 5 W "off" mode: 1 W	"on" mode: depends on size
Tumble dryer					EU energy label A or B end of cycle: 5 W "off" mode: 1 W	

* *GEA: Group for Efficient Appliances*
** Equal to Energy Star
*** Equal to Group for Efficient Appliances
Sources: www.europa.eu.int/comm/environment/ecolabel, www.blauer-engel.de, www.ecolabel.no, www.gealabel.org.

Table 4.2 Illustration of policy coverage

Products (in descending order of power consumption)	EACEM (VA)	EU (VA)	GEA (VA/ labelling/info)	Switzerland (VA/info)	UKMTP (info)	Energy Star (VA/label)	Japan Electronics Mfc. Industry (VA)	Top Runner (standard)
	Existing policies to address standby power consumption							
			(AU, DK, FI, F, G, N, SW, CH)*	(CH)	(UK)	(US, J, Aus, NZ)	(J)	(J)
Telephone					X (Tel. chargers)			
Television	X	X	X	X	X	X	X	X
VCR	X	X	X	X	X	X	X	X
TV-VCR combination		X	X	X		X	X	
Office PC and monitor			X	X	X	X		X
Copier			X	X	X	X		X
Complete hi-fi system	X		X	X	X	X	X	
CD and DVD player	X		X	X	X	X	X	X
Office machine			X	X	X	X		
Office electronic typewriter								
Microwave oven								
Analog cable box								
Residential PC and monitor			X	X	X	X		
Residential answering machine								
Range (oven/cooker)					X			
Hi-fi amplifier	X		X	X	X	X	X	
Office matrix printer			X	X	X	X		

Products (in descending order of power consumption)	EACEM (VA)	EU (VA)	GEA (VA/ labelling/info)	Switzerland (VA/info)	UKMTP (info)	Energy Star (VA/label)	Japan Electronics Mfc. Industry (VA)	Top Runner (standard)
			(AU, DK, FI, F, G, N, SW, CH)*	(CH)	(UK)	(US, J, Aus, NZ)	(J)	(J)
Clock radio								
Tumble dryer								
Cordless phone								
Office laser printer								
Residential ink jet printer								
Battery charger								
Residential machine								
Aerial amplifier (analog satellite receiver)								
Residential laser printer								
Residential electronic typewriter								
Garage door opener								
Mobile phone								
Audio portable								
Doorbell transformer								
Power speaker								
Induction cook top								
Office ink jet printer								
Office answering machine								

(continued)

Products (in descending order of power consumption)	EACEM (VA)	EU (VA)	GEA (VA/ labelling/ info)	Switzerland (VA/ info)	UKMTP (info)	Energy Star (VA/ label)	Japan Electronics Mfc Industry (VA)	TopRunner (standard)
			(AU, DK, FI, F, G, N, SW, CH)*	(CH)	(UK)	(US, J, Aus, NZ)	(J)	(J)
Plug-in power supply unit (including chargers, wall packs)								
Office notebook								
Residential matrix printer								
Workstation – dumb terminal								
Cassette recorder								
Residential notebook								
Thermotransfer printer								
Scanner								
Integrated receiver decoder					Digital decoder			
Energy-saving device								
Multifunctional device								
Lighting products								
Clothes washer								
Dishwasher								
Refrigerator/freezer								
Central heating								
Air-conditioning								

AU = Austria, DK = Denmark, FI = Finland, F = France, G = Germany, N = Netherlands, SW = Sweden, CH = Switzerland, US = United States, J = Japan, Aus = Australia, NZ = New Zealand.

**not all the participating countries have adopted the GEA scheme for IT equipment (though all have adopted the scheme for Consumer Electronics).*

POLICY TOOLS

INTRODUCTION

Transforming markets to improve energy efficiency and reduce standby power consumption is a complex task that requires a variety of policies. Several approaches require government decisions and legislation. Some approaches rely on voluntary action, with a minimum of government intervention; an example is private sector procurement practices. Others require co-operation or negotiation among various stakeholders, particularly between industry and government. Policy objectives range from removing the least efficient products from the market to encouraging products with the lowest power consumption levels, to development of new technology. The

Figure 5.1 **Classification of policy mechanisms to improve appliance energy efficiency**

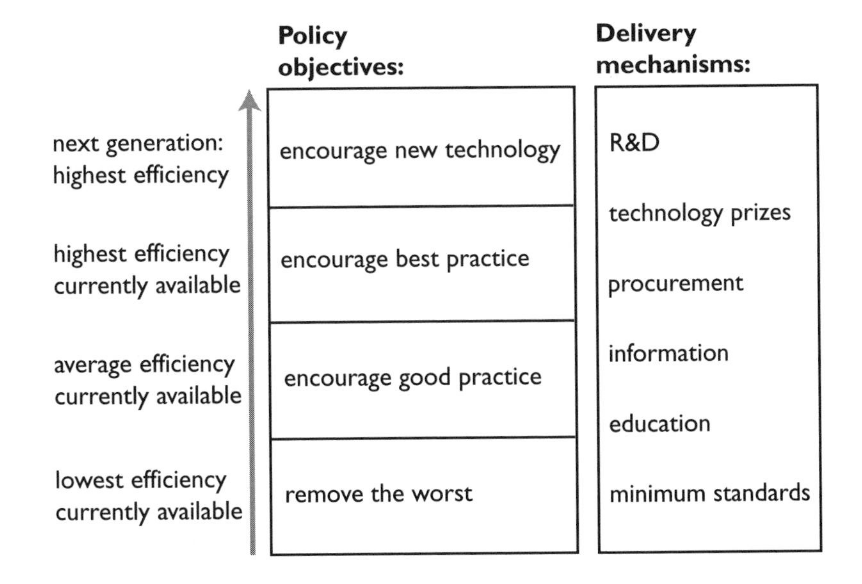

tools range from minimum standards or specifications to information and education, procurement policies, and research and development. Packages of policy tools are needed to address standby power in a comprehensive and effective way. No one tool can solve the problem alone.

Figure 5.1 illustrates how these objectives, approaches and policy tools can combine to tackle different aspects of the standby losses problem.

STANDARDS

Introduction

Energy efficiency standards for appliances and equipment play an important role in OECD countries' strategies to meet energy and environmental policy goals. In most OECD countries, energy efficiency standards are already in place for products that consume a lot of power in active mode, such as refrigerators, air-conditioners and water heaters. These standards address the largest quantity of the energy consumed by these products, but some do not address standby power.

Standards are typically established in regulations, but it is not always necessary to regulate to create standards. Industry "norms" can be developed without any government involvement. Voluntary industry standards are often later incorporated into government regulation. In most cases, this is not necessary, as voluntary industry standards are universally accepted by the industry sector. Non-binding technical specifications can be drawn up by industry or by an international body, such as the International Electrotechnical Commission (IEC). Most manufacturers follow such standards, because they prefer to have one standard for their products wherever they are sold. At present, however, only a few standards for energy efficiency include criteria for measurement of standby energy consumption.

Effectiveness of standards as a policy for addressing standby power

Energy efficiency standards ensure that efficiency is incorporated into product design. Many consumers are not aware of, or not concerned about, energy efficiency. When making a purchase decision, consumers often place a premium on other factors, such as the product's features, overall performance or price. Product characteristics such as low noise, size or speed play a much greater role in customer preference than energy efficiency. As a result, these characteristics are the primary focus for manufacturers in both product design and marketing strategies.

Including standby power in energy efficiency standards is an effective option for addressing this issue, particularly as it is unlikely to affect product performance. The effectiveness of standards as a policy tool for standby power depends on how they are designed and implemented. The most effective standards require a certain level of energy efficiency or set a maximum power level in standby mode, but do not dictate how this level should be achieved. Some standards (such as the Corporate Average Fuel Efficiency standards for vehicles in the United States) allow manufacturers to produce some products with lower efficiency and others with higher efficiency, so that on average their product stock meets the standard. The Japanese Top Runner programme allows manufacturers to achieve the targets on the basis of a sales-weighted average of all their products. Criteria for standards can include a cost-effectiveness feature to ensure that they do not lead to significant cost increases. One concept is "best available techniques not entailing excessive costs" (BATNEEC).

The time allowed between announcement of standards and their implementation can greatly affect their acceptability and effectiveness. If tough standards are implemented too quickly, they will impose unacceptable costs on manufacturers, who will be unable to sell their lower-standard stocks and adapt their product lines before the standard comes into force. If new laws are needed to support regulations, they can be complicated and time-consuming to implement. If the transition period is too long, however, the standard may be obsolete before it comes into force. These problems are particularly relevant to standby power standards, as the rapid rate of technological innovation can, within a short period of time, completely change the achievable standard.

Monitoring and enforcement costs need to be considered. Appliances that consume large amounts of power in standby modes are increasingly numerous. So a large number of products would have to be regulated and monitored. The complex technical nature of energy efficiency standards for standby power also affects the cost of test protocols and measurement techniques. If these do not already exist, it can take much time and effort to develop and implement them.

Scope for international approaches

Harmonised international standards or norms would reduce costs and enhance international trade. International standards reduce measurement and certification costs. They also help avoid the confusion of conflicting claims from manufacturers using different test protocols.

Trade and technology transfer are much easier among countries with similar standards. Larger markets allow economies of scale and lower prices for efficient products and technologies. This, in turn, increases manufacturers' incentive to develop them. Harmonised standards would therefore enlarge the energy-efficient segments of product markets for both products and component technologies. If standards are not harmonised, they can form barriers to trade.

VOLUNTARY APPROACHES: WORKING BY CONSENSUS

Introduction

Voluntary agreements (VA) between industry and government are an increasingly common policy tool. Most existing policies on standby power are voluntary in nature.Voluntary approaches can be informal agreements, with no guarantee they will be met and no penalties for failure to meet them. But they can also be negotiated agreements with legally-binding targets (where the government can impose penalties if the agreed targets are not met).[20] Voluntary approaches have been used for labelling as well as setting targets for maximum power levels in standby modes.

Monitoring and reporting are elements in most voluntary agreements. This in itself can raise awareness and provide information that makes it easier to tackle the problem. Some voluntary agreements contain a mandatory monitoring and reporting element.[21] In some cases, a third-party organisation is responsible for monitoring compliance. In others, industries are encouraged to report the basis of self-assessments. Some voluntary agreements rely in part on competition to help monitor compliance, as companies scrutinise one another's behaviour in the hope of gaining a competitive advantage.

Effectiveness as a policy for addressing standby power

Voluntary agreements enable industry to negotiate realistic goals and timetables for meeting them. The voluntary agreements that have been implemented to date have been set at a level which the

20. *Voluntary Agreements with Industry (OECD internal document, 1997).*
21. *For example the US Climate Challenge programme and the Dutch Long Term Agreements.*

industry considers feasible and cost-effective. In some cases, the voluntary targets have been met or exceeded well within the agreed time-frame. Because it is difficult to define what would have occurred in the absence of such agreements, it is not possible to judge their effectiveness with precision. But they have clearly proven an effective and flexible instrument in many countries, particularly in minimising industry compliance costs.

Voluntary agreements are a particularly useful tool in well-organised sectors with a small number of large players or where there are representative industry associations that can negotiate on behalf of their members. They can be very effective where a strong relationship exists between industry and government, where monitoring systems are in place and where the negotiations are not too lengthy.

Scope for international approaches

International voluntary approaches have already generated widespread industry support. As with international standards, it is necessary to overcome differences in test protocols and measurement techniques.

ECONOMIC INSTRUMENTS: MAKING THE MOST OF MARKET INCENTIVES

Introduction

Economic instruments include energy taxes, tax credits, tradable permits, fees and rebates. Energy taxes raise the cost of energy. This provides an incentive for households and offices to save energy by purchasing more efficient appliances. Energy taxes can improve product efficiency if demand for efficient appliances increases and manufacturers respond by changing product design.

Tax credits can achieve the same objective through direct financial incentives to consumers and businesses. Tradable permit systems could also be used to give manufacturers flexibility in meeting standards.

Fees and rebates operate in similar ways. Fees raise the cost of less efficient products. Rebates reward the customer for buying more efficient products. These two instruments can be combined in "feebate" schemes, where fees that are paid for inefficient products are used to finance rebates for efficient products.

One problem with economic instruments is that few consumers make buying decisions on the basis of cost alone. Most consider several other features of the desired product.

Effectiveness as a policy for addressing standby power

Taxes and tax credits

The impact of energy taxes will depend on consumers' response to increases in their energy bills. However, energy taxes are far less likely than other policy tools to have a direct or a major impact on product energy efficiency.

A tax exemption measure that has been enacted by the State of Maryland, in the United States, provides a promising approach for influencing consumer demand for efficient products. The "Maryland Clean Energy Incentives Act" waives the sales tax on some Energy Star appliances. This reduces the cost of these appliances to consumers by 5 per cent.[22] This is a direct financial incentive for consumers to buy Energy Star appliances. Exempting efficient appliances from value-added taxes in other countries, particularly those with high tax rates, could encourage sales.

22. *www.energystar.gov/mdact/synopsis.pdf*

Tradable permits

Tradable permit systems could be one way for manufacturers to achieve an agreed degree of energy efficiency. Tradable permits have the added advantage of flexibility. Manufacturers who find it difficult to meet energy efficiency requirements could buy permits from manufacturers who do better than the targeted level. But the infrastructure needed to monitor enforcement and trading would be more complex than for other economic instruments and therefore more costly to administer.

Fees and rebates

Fees and rebates could be a workable tool for addressing standby power. There are already some examples of rebate systems, although it is too early to judge their effectiveness. One example is a proposal by the US Department of Energy for a rebate scheme to transform the market for wall packs (external power supplies). Many electric utilities in the United States offered rebates on energy-efficient appliances in the 1990s. In Germany, some utilities already have a rebate scheme for energy-efficient consumer electronics, and others are planning similar schemes.

Scope for international approaches

Attempts to implement internationally harmonised taxation are very controversial and none has succeeded to date. The lengthy negotiations over harmonised energy taxes in the European Union show that international harmonisation of energy taxes is fraught with difficulties. Less well integrated countries would face even greater obstacles to the harmonisation of taxes.

Tradable permits would only work as an international system if governments agreed to allow their manufacturers to purchase from abroad the right to manufacture less efficient products, and if energy efficiency standards, test protocols, monitoring and measurement methods were harmonised. It is hard to imagine an international fee, rebate or feebate system unless energy efficiency criteria, test and measurement methods were agreed upon at the international level.

INFORMATION AND EDUCATION: A USEFUL POLICY COMPLEMENT

Introduction

Awareness-raising programmes

Information and education to raise consumer awareness are often an important part of a package of measures to improve energy efficiency. Information, education and training measures generally target behaviour. They include funding education at various levels and orienting the curriculum to include information on the importance of energy efficiency. Other approaches are advertising on television, radio and web-sites, and in newspapers, to try to influence behaviour.

Labelling

If they are readily understandable, energy labels can provide the information needed to make informed purchase decisions. Labels also raise the visibility of energy efficiency as an issue. Appliance labelling can be a convenient means to help monitor the market and assess efforts to move buyers towards more energy-efficient products. Even though its primary objective is to influence consumer purchase decisions, labelling also can help in compiling information on market transformation.

Databases

Databases are a useful way to provide information to consumers, particularly now that millions of consumers can access them on the Internet. There are already quite a few databases that provide information on the power consumption of appliances in standby modes. There are well over a thousand appliances listed in the GEA database (www.gealabel.org). Visitors to the web-site can view a list of appliances sorted by appliance type and country. An Internet database also provides information on all Energy 2000 products under the Swiss counterpart of the GEA label. The US Energy Star programme (www.energystar.gov) and the UK Market Transformation Programme (www.mtprog.com) also offer detailed databases of products. These help manufacturers of efficient products to market them, make it much easier for consumers with Internet access to find efficient products and reassure consumers that the manufacturers' claims are legitimate.

Effectiveness as a policy for addressing standby power

For products such as VCRs, there is limited potential for energy savings through changes in behaviour because they have no "off" switch. For other products, persuading people to switch off electrical appliances rather than leave them in standby mode could save energy. Switching off televisions and hi-fis at the set rather than leaving them in standby mode could reduce energy consumption significantly. The consumer can also reduce the total energy consumption of power-supply units by disconnecting them from the wall socket when they are not in use. But many people find it inconvenient to switch off appliances and they do not realise the costs of standby power. Therefore, measures that target behaviour alone have limited impact in the short term.[23]

23. www.mtprog.com

Even with labelling and increased awareness of how improved energy efficiency can reduce their electricity bills, it can be very difficult for consumers to distinguish between efficient and inefficient appliances. Many people do not make the effort. Although some labels, such as Energy Star, the EU-eco label and the German Blue Angel, cover a large portion of the market, many other eco and energy labels cover only a small fraction of the appliances sold. There can be multiple competing labels, a situation that tends to be more confusing than informative. Another difficulty with energy labels is that product development can be so rapid that any new standard may become out of date very quickly.

PROCUREMENT PROGRAMMES: AN INSTRUMENT TO MAKE A DIFFERENCE

Introduction

Technology procurement programmes encourage innovation by promising to purchase very efficient products. Buyers concentrate their buying "power" in a single specification for an innovative product. The company that meets these specifications knows that it can sell the product. The publicity that accompanies procurement programmes increases the visibility of features of the technology such as energy efficiency.

Effectiveness as a policy for addressing standby power

Procurement programmes can greatly increase the effectiveness of voluntary standards and labelling measures. Procurement programmes by large institutions and private companies provide strong incentives for developing energy-efficient products by increasing the size of bulk purchase contracts. In principle, market demand from large companies can increase production runs of energy-efficient technologies and drive costs down.

In practice, however, technology procurement projects have proved difficult to carry out. Since technology procurement aims at significant innovations, the period between the first contacts with the buyers and the delivery of the product can be several years. This leaves the buyers uncertain over when and whether they will get the product they want. Buyers are reluctant to commit themselves to specifications and purchases when they will see the effect only some years in the future.

Another difficulty is that technology procurement is often related to specific products. But some of these products may be used together with other products, or may be one part of a system. Often the problem is not a purely technical one of increasing the energy efficiency of a given product, but one of successful introduction of the product to the market. Technology procurement addresses only the former.

Although technology procurement is not an easy instrument to use, it can stimulate constructive discussion with manufacturers. It can change attitudes and draw designers' attention to energy issues.

Scope for international approaches

Although many product markets are international or even global, international procurement presents particular problems. The first difficulty is the extra effort that is needed to co-ordinate the activities of all parties in the participating countries. Some governments have a national focus and aim only to stimulate domestic industry. In such cases, they may not be interested in benefiting foreign companies that are competitors of national manufacturers. However, multinational companies can more easily participate if a procurement programme is international in scope. For certain appliances, the same specifications can be used in multiple countries, which reduces manufacturer and customer costs.

NEXT STEPS FOR POLICY-MAKERS

INTRODUCTION

Standby energy cannot be viewed in isolation. Some energy consumption in standby modes is inevitable and supports functions important to consumers, such as remote control. In other cases, standby power fuels safety or security features. Because standby modes themselves are much less energy-intensive than active modes, they are an essential and important tool in reducing the overall energy consumption of many household appliances and office equipment. Nevertheless, there is a consensus that standby power modes are less efficient than they could be and are not yet adequately addressed in energy efficiency policies.

Standby power consumption provides a good opportunity for reducing both energy consumption and greenhouse gas emissions. Through co-operation among governments, industry and consumers, and the co-ordination of international policies, standby modes can be made more efficient, thereby reducing the overall demand for power. Where it is cost-effective, product design should automatically include low-power standby modes and should minimise energy consumption in standby modes. Existing engineering resources allow for substantial low-cost reductions in standby power consumption. In many cases these can be achieved without affecting the performance of the product.

Given the continuing rapid changes in technology, including the growing prevalence of networked homes and offices, it is critical that countries move forward as soon as possible. A co-operative, co-ordinated effort can help ensure that reducing standby power consumption becomes an integral part of product research and design. Power management should be transparent to the consumer and should occur automatically, without inconvenience to the user and with minimal impact on the effectiveness of the appliance.

MEDIUM-TERM PROSPECT

The IEA hopes that a co-operative, co-ordinated effort to reduce standby power consumption could, by 2010, result in energy efficiency being incorporated automatically into the design of appliances. It will be normal to minimise standby power consumption and to use standby power modes to optimise overall appliance efficiency.

THE CHALLENGE AHEAD

The challenge facing IEA Members is to achieve greater energy efficiency through improved product design and, to the extent possible, change consumer preferences. The difficulties stem from several characteristics of products that consume power in standby modes:

- In most cases, the amount of standby energy consumption in individual appliances is small, but the large numbers of appliances lead to high energy losses.
- Consumers often are not aware of, or concerned about, standby power consumption.
- Standby power consumption is influenced by many different parties in addition to the manufacturer. For example, service providers influence the standby power consumption of networked appliances.
- Many appliances with standby modes have low profit margins and short product lives, owing to the introduction of newer, more advanced models within short time periods.

These characteristics need to be taken into account when developing policies.

GUIDING PRINCIPLES

The active support and commitment of all parties is needed to reduce standby power. The IEA suggests the following principles to guide the development of effective policies on standby consumption:

- All parties, including industry and consumer groups, should be encouraged to participate in policy deliberations and development. This is both to ensure that the policies are well designed and realistic, and to secure the participants' full co-operation. Regular discussions with manufacturers and other relevant parties would help policy-makers foresee future developments, including ways to reduce the effects of standby power consumption.
- Policy objectives should be expressed in terms of overall goals and desired outcomes. They should allow for and encourage technological advance and innovation.
- Policy-makers should avoid trying to define standby power and instead focus on ways to improve efficiency. Different definitions of standby power create confusion and there is no single correct definition.
- Policies should be non-discriminatory regarding technology and should refrain from embracing programmes or criteria that impede international trade.
- Because the product market is changing so rapidly, policies should be very flexible. Policies should encourage further innovation to minimise or optimise standby power consumption and ensure a minimum efficiency level for such consumption.
- Approaches to reducing standby power consumption should be international. National policies should be harmonised to minimise administrative burdens and associated costs, to lever

national and regional promotional investments and to minimise the risk of non-tariff trade barriers.

- Test protocols are the basis for any policy on standby power, and harmonised test protocols are a prerequisite for a global approach to the problem. Standby power should be included in all energy test protocols that address appliances with relatively high standby power consumption.

A CALL FOR INTERNATIONAL COLLABORATION

Introduction

Many of the appliances that consume standby power are internationally traded. So standby power is an ideal candidate for international, co-ordinated action. Improving energy efficiency is one of the most cost-effective ways to reduce greenhouse gas emissions from energy. The low costs of dealing with standby power and the benefits from energy savings and emissions reductions would be spread over all countries.

Standby power consumption is already being addressed in many IEA countries, as discussed in Chapter 4. But the programmes that have been implemented to date are not globally consistent. Each has its own definitions, methods, criteria and test procedures. Different programmes with different sets of rules create barriers for manufacturers who have to comply with them.

A successful international approach to standby power would build on existing policies and processes. In addition to the national and regional policies that were presented in Chapter 5, there are many international forums and programmes that could be used to deal with standby power. Table 6.1 summarises some of the most relevant international programmes.

Table 6.1 **Summary of international programmes**

International organisation	Main programmes
APEC	Asia Pacific Economic Co-operation ■ common action on standards and test protocols
CTI	Climate Technology Initiative ■ technology assessment, analysis and strategy working group (Co-operative Technology Implementation Plans) ■ technology awards ■ training (e.g. energy efficiency workshops)
IEA	International Energy Agency ■ analyses and publications ■ exchange of information ■ co-operative agreements covering energy efficiency market transformation activities (information exchange, research studies, specific projects)
IEC	International Electrotechnical Commission ■ electrotechnical and information technology standardisation ■ technology awards
WBCSD	World Business Council for Sustainable Development ■ contributing a business perspective to policy development through research; research on resource efficiency, including development of indicators and monitoring and reporting guidelines
UNFCCC	United Nations Framework Convention on Climate Change ■ inter-governmental agreement to limit greenhouse gas concentrations in the atmosphere to "safe" levels ■ Kyoto Protocol to the UNFCCC, which places legally-binding limits on IEA countries' emissions

IEA forums and co-operative agreements

The IEA is the main international forum on energy issues for developed countries and is well placed to facilitate an international approach on standby power. The IEA sponsors a wide range of targeted collaborative activities. Examples include co-operative agreements for technology R&D and market transformation, which are referred to as Implementing Agreements, and the Oil Markets and Emergency Preparedness processes. The IEA was the primary sponsor of the two international workshops and three task forces that led to this publication.

IEA forums could provide co-ordination among existing schemes and opportunities for discussion of items of common interest, such as future developments in networked appliances and power management. Member countries could develop a road map towards a single international labelling scheme, aiming for ambitious criteria and the application of power management. IEA activities might include producing newsletters and electronic bulletin boards to facilitate information exchange; promoting networking among individuals and organisations; hosting *ad hoc* working groups, workshops and conferences; and co-sponsoring a journal on energy efficiency.

IEA legal frameworks for international collaboration may also provide effective vehicles for addressing standby power. The IEA Implementing Agreement on Demand-Side Management, for example, already deals with energy efficiency research. Annex III of that agreement focuses on market transformation activities that aim to create demand for (and supply of) energy-efficient products. Standby power consumption is not yet considered explicitly in Annex III, but discussions are under way to establish a sub-task on standby power within this annex.

Other IEA-sponsored or co-ordinated products might include research studies on market transformation and technology issues;

reports or case studies on the various programmes and policy approaches to standby power issues; and comparative studies of efficiency labelling and energy testing and rating methods.

Given the IEA's expertise on global energy matters, and its excellent reputation for consensus-building, it is well positioned to play a key role in achieving progress on reducing standby power consumption.

CONCLUSION AND RECOMMENDATIONS

Conclusion

An international approach to reduce standby power makes sense. The problem is real, significant and should no longer be ignored. It exists in most OECD countries. Economic activities are becoming increasingly global, especially for end-use technologies such as electric equipment and appliances that consume power in standby modes.

There are enough energy-efficient solutions already available on the market to make substantial reduction of standby power consumption a realistic objective. In a business-as-usual scenario, some of the low-standby solutions would no doubt be brought to market but not fast enough to compensate for the overwhelming growth of standby power consumption that will come with the next generation of electronic equipment. Tackling the problem from an international platform is the best way to reach the necessary scale to increase the penetration of these technologies in a global market.

Reducing standby power at the international level is possible. Indeed it is already happening. Multinational companies have understood the need to improve the energy efficiency of the standby mode of the equipment they sell. This is encouraging, but government intervention can stimulate and reinforce such achievements.

The IEA standby power project has opened the door to an international policy co-operation among policy-makers.

There are many technologies that could benefit from co-ordinated international effort. Examples include some lighting technologies, transformers, Internet machines and electric drives.

Recommendations for immediate follow-up activities

1. Ensure support and participation in IEC TC59 on standby power

Standby power should be included in all energy test protocols and in all energy efficiency policies on products with significant standby power requirements.

The International Electrotechnical Commission (IEC) Technical Committee TC59 is responsible for household appliances. In October 1999, TC59 created an *ad hoc* working group to examine test procedures for standby power on appliances and electric equipment. The initial tasks are to:

- Collect general information on programmes that require measurement of standby energy consumption.
- Document methods currently in use.
- Examine technical issues associated with the measurement of standby power (wave shapes, power factor) and its applicability.
- Deal with issues such as test conditions, methodologies, instrumentation and accuracy for standby power.
- Assess the need for a horizontal product standard in light of the data collected.

The TC59 standby power working group does *not* cover:

- Developing programmes for influencing standby power consumption.
- Developing labels or markings associated with standby power.
- Setting limits for standby power consumption.

Countries are encouraged to support and participate in the new IEC TC59 working group on standby power. They are also encouraged to support TC74 and TC92, which deal with information technology and consumer electronics, respectively.

2. Establish an international voluntary programme

The global and dynamic nature of the market for appliances with standby modes will be best served by co-ordinated efforts among industries and governments. Such efforts could take place in an international working group with representatives from both industry and government. Experts participating in the working group can keep the business community up to date on the representation of labels and the prevailing state of technologies in the market-place.

Internationally co-ordinated efforts would reduce the burden on manufacturers of global products, thereby encouraging their co-operation with, and support for, greater reductions in standby power consumption. Perhaps more importantly, an international approach would eliminate the confusion created by redundant energy efficiency labels and labelling schemes. It could simplify the process of educating and informing consumers about the issue, and stimulate greater demand for energy-efficient products and appliances.

The Energy Star logo is now well known in many countries. Given its growing acceptance and recognition in markets throughout the world, it presents an excellent platform for addressing standby power issues, which additional countries may wish to adopt.

The Energy Star programme was initiated in the early 1990s by the US Department of Energy and the US Environmental Protection Agency (EPA), as described in Chapter 4. Some IEA Member countries have already adopted the Energy Star labels. In 2000, the US EPA and the European Commission were finalising negotiations on a co-ordinated use of the Energy Star logo on Information Technology. Building on the success of the Energy Star label, and strengthening it, could be an important step towards reducing standby power consumption in information technology and some residential equipment.

The agreement between the US EPA and the European Commission on the Energy Star logo for office equipment presents a basis for a wider co-ordinated use of the scheme by more IEA Member countries and for covering more equipment. When properly established, an international Energy Star scheme can embrace many programmes to reduce standby power.

An international Energy Star programme should also involve countries that are not IEA members and work with other relevant international bodies, such as the World Trade Organization (WTO), the IEC and the International Organization for Standardization (ISO). The programme could establish international norms, testing protocols and monitoring processes, in collaboration with the IEC and other relevant bodies.

The voluntary agreements negotiated between the European Commission and some individual European appliance industries (see Chapter 4) can be expanded to other IEA Member countries. The standby power levels set in these voluntary agreements should also be co-ordinated with the Energy Star scheme described above.

For numerous appliances, a combination of an Energy Star standby power threshold and a voluntary agreement between government and the industry is the most appropriate measure available to reduce standby power.

The recent European agreement to reduce the standby power modes of wall packs (external power supplies) could easily be duplicated and implemented in other regions of the OECD.

To the greatest extent practicable, international voluntary programmes to reduce standby power should take into account, and be harmonised with, existing regulatory schemes, such as Japan's Top Runner programme. In general, international voluntary programmes should include consideration of national circumstances and should have more stringent targets than national mandatory regulations. International voluntary programmes should take into account the situations of major manufacturing and consuming countries.

3. Develop guidelines for lowering standby use in appliances not covered by any programme

Many of the newest technologies will require some standby power. The "networked home" may potentially be a high-standby home. It is therefore important to develop guidelines for lowering standby power use in new appliances and appliances not currently covered by any programme.

4. Avoid the proliferation of labels to reduce standby power

Some regions or countries have introduced their own label or scheme to encourage the purchase of equipment with low standby power consumption. Should such labels be maintained despite the introduction of a more international scheme, it would be appropriate at least to ensure that the criteria to reduce standby power consumption converge.

5. Address the specific case of set-top boxes for digital television

Television broadcasting is rapidly moving towards digital technology. Set-top boxes are likely soon to represent a significant new standby power demand in most economies. Countries should rapidly co-ordinate their efforts to develop future communication protocols, and their other efforts to deal with the new generation of set-top boxes, to ensure that the standby power mode is as energy-efficient as possible. Service providers have to be closely associated with this work.

6. Include standby power information in existing appliance energy labels

Appliance energy labels exist in most IEA Member countries. In a large majority of them, there is no indication of the energy consumed while the appliance is in standby mode. It may be appropriate to include in forthcoming updates of appliance energy labels an indication of the standby power consumption.

7. Stimulate research on new low-standby technologies

New solutions to reduce standby power should be encouraged. IEA Member countries should consider assisting the research and development activities of manufacturers encountering technical obstacles to reducing standby power.

ANNEX 1 DEFINITION OF STANDBY POWER

INTRODUCTION

There are many definitions of standby power use. Some describe it in functional terms, such as "the power consumed by an appliance when switched off or not performing its primary functions". Some implicitly limit their definitions to electronic products, while others include pilot lights in natural gas appliances, storage losses in water heaters, and certain features of refrigerators that constantly draw power. Others have tried to use a technically simple definition, such as "the minimum power consumption of a device while connected to the mains". There is no single correct definition, but the diversity of definitions has been a major obstacle to making a global estimate of standby power consumption.

In 1999, the International Energy Agency created a task force of representatives from industry, government and technical bodies to develop a definition of standby power that would be suitable for estimating the size of the problem and to develop policy options and further collaborative action. The consensus definition[24] is rather complex, and it depends upon the device being measured.[25]

The task force recognised that no definition would satisfy everyone. Therefore, the task force sought to develop definitions that would facilitate further action on standby power. It was also agreed that the definition need not be technically rigorous and need not cover all possible situations, but that it should be comprehensive enough to facilitate subsequent policy discussions.

24. IEA, 2000.

25. For practical reasons, most of the measurements in this study were taken when the devices were at their lowest electrical power consumption while connected to the mains.

The task force developed the following definition of standby power, along with clarifying remarks and observations.

STANDBY POWER DEFINITION

Standby power use depends on the product being analysed. At a minimum, standby power includes power used while the product is performing no function. For many products, standby power is the lowest power used while performing at least one function. This definition covers electrical products that are typically connected to the mains all of the time.

Based on this definition, certain types of products generally do not have standby power consumption. This includes, for example, products that have only two distinct conditions: "on" and "off", where the product does not consume power when it is switched off.

This definition is not intended to supersede or replace the definition underlying existing programmes.

Standby power consumption should be viewed in the context of the overall efficiency and performance of the product.

Specific measurement procedures should be determined by the appropriate standardisation bodies. It is recommended that these bodies consider the role of power factor and power management, and choose a suitable measurement interval.

For electrical products, standby power and standby energy consumption are generally expressed in average watts and kilowatt-hours, respectively.

The task force discussed the most appropriate word for "standby" energy consumption. The underlying problem is that this energy consumption covers several different product conditions. In the end, the task force used "standby power". The appropriate technical bodies may choose to develop more precise terms that are consistent with the definition and test procedures used.

The phrase "no function" includes the situation in which the secondary load is disconnected from the power supply. This is often called the "no load" power use.

ANNEX 2
ASSESSING CO_2 EMISSIONS FROM STANDBY POWER

ABSTRACT

Studies have shown that standby power is responsible for 20 to 60 watts per home in developed countries. Standby power accounts for about 2 per cent of OECD countries' total electricity consumption, while power generation to support standby modes produces almost 1 per cent of those countries' electricity-related CO_2 emissions. Replacement of existing appliances with appliances having the lowest standby power use would reduce total standby power consumption by over 70 per cent. This represents close to 4 per cent of the difference between projected OECD emissions and the CO_2 emissions levels that would meet the Kyoto targets. Other strategies may cut more carbon emissions, but standby power is unique in that the reductions are best accomplished through international collaboration. Both modest costs and large benefits would be spread over all countries.

INTRODUCTION

Standby power has recently gained recognition as a unique and significant use of electricity. A recent study[26] of video-cassette recorders (VCRs) in the United States showed that more electricity is consumed when VCRs are in the standby mode than when actively recording or playing. A study in New Zealand[27] revealed that more than 40 per cent of microwave ovens consume

26. *Rosen and Meier, 2000.*
27. *EECA, 1999.*

more electricity in standby mode, by powering the clock and keypad, than in cooking food. Many countries have already begun programmes to reduce standby power in the most prominent appliances, such as televisions, VCRs and audio equipment.

Recent investigations and calculations suggest that standby power is a more pervasive and a larger problem than was first thought. Field measurements in a few homes scattered around the world demonstrated that standby power consumed between 3 and 13 per cent of those homes' total electricity. Standby power consumption also appears to be growing rapidly, as more appliances are built with features that lead to standby power consumption.

Before starting a vigorous international programme to reduce standby power consumption, it is essential to know how much power is currently being used and how much might be saved by prompt actions. Does the size of the problem justify international co-operation? If so, how much can be achieved compared with other mitigation measures? This annex addresses the first question, that is, "how much electricity does standby power consume globally?" With this information, standby power's contribution to global CO_2 emissions can be estimated.

EXISTING ESTIMATES OF RESIDENTIAL STANDBY POWER IN INDIVIDUAL COUNTRIES

One perspective on global standby power consumption can be obtained by compiling estimates from individual countries. The approaches used by researchers to make the estimates vary greatly, but fall into two categories: "field studies" and "bottom-up" estimates. These are described briefly below.

Field studies have been conducted in France, Japan, Australia, the United Kingdom and New Zealand. The estimates were based on a group of carefully monitored homes. The French field study[28] was one of the largest end-use studies in the world, and the largest single compilation of standby power measurements. The sample of 178 households around France is believed to be representative of the entire housing stock, at least regarding average penetration of specific types of equipment and average electricity consumption. Despite the extreme care taken by the researchers while on site, the standby power consumption of some appliances, such as doorbells, could not be checked and recorded. In other cases, the researchers simply overlooked appliances with standby power consumption as it is not easy to identify the standby functions of many appliances.

Table A2.1 presents the standby power consumption of appliances found in the 178 French households surveyed in 1998 and 1999. The table illustrates the diversity of the equipment found with standby power uses corresponding to the definition given in Annex 1. More than 70 categories of appliances are listed. The survey shows the range of standby power use within each category of equipment. The average household standby power use ranges from 29 watts to a possible 38 watts (depending on assumptions about the time of use of the standby mode for some appliances). Standby power use in individual homes ranged from 1 to 106 watts.

28. Sidler, 2000.

Table A2.1 Survey of the standby power mode for video, hi-fi, and information equipment found in 178 French households (1998-99)

Category	Appliances	Maximum standby power (watts)	Minimum standby power (watts)	Average standby power (watts)	Number of models monitored
Video	Television	22	1	7.3	205
	Individual TV amplifier	4	1	1.8	33
	VCR	30	1	9.9	169
	Video-cassette rewinder	1	1	1.0	3
	CDV player	5	5	5.0	1
	DVD player	15	15	15.0	1
	Infrared headphone	5	1	2.1	8
	DSP sound system	1	1	1.0	1
	Pro logic speaker	2	2	2.0	1
	Infrared speaker	9	9	9.0	1
	Video games	7	1	1.7	20
	Hertz TV decoder	16	9	11.0	34
	Satellite dish decoder	17	5	8.7	26
	Cable TV decoder	23	3	9.5	4
	UHF connector	10	10	10.0	1
Hi-fi	Hi-fi stereo	24	1	7.2	108
	Hi-fi amplifier	9	1	4.0	5
	Vinyl disk player	1	1	1.0	3
	Tuner	5	1	2.4	7
	CD player	7	1	3.1	18
	Cassette player	6	1	2.2	6
	Portable CD player	1	1	1.0	1
	Radio / cassette tape player	4	1	1.7	41
	Radio / alarm clock	4	1	1.4	175
	Alarm clock	2	1	1.7	3
	Miscellaneous hi-fi TV / video	34	4	14.4	8

Category	Appliances	Maximum standby power (watts)	Minimum standby power (watts)	Average standby power (watts)	Number of models monitored
Information technology	PC central unit	2	2	2.0	2
	PC monitor	10	1	6.5	4
	PC central unit + monitor	3	2	2.7	3
	PC whole unit	27	1	6.9	14
	Tension stabiliser	18	14	15.7	3
	Laptop PC	20	1	6.5	4
	Modem	6	3	4.3	3
	Ink jet printer	8	1	3.8	13
	Laser printer	4	4	4.0	2
	Scanner	6	5	5.5	2
	PC speaker	5	1	3.0	2
	Desk calculator	2	2	2.0	1
	Typewriter	5	5	5.0	1
	Photocopy machine	10	10	10.0	1
Telephone	Cordless telephone	7	1	2.6	100
	Independent answering machine	6	1	2.8	56
	Combined phone-answering machine	11	1	5.1	31
	Interphone	3	3	3.0	1
	Minitel (basic Internet access unit)	9	5	6.2	11
	ADSL connection box	3	3	3.0	1
Cooking	Coffee maker	1.5	1	1.1	7
	Induction cooktop	18	4	13.2	10
	Microwave oven	12	1	3.5	32
	Kitchen oven	18	6	14.5	4
	Battery recharger	2	1	1.3	4
	Musical keyboard	3	1	1.8	4
	Electric fence	1	1	1.0	1
	Security system	1	1	1.0	2

(continued)

Category	Appliances	Maximum standby power (watts)	Minimum standby power (watts)	Average standby power (watts)	Number of models monitored
Miscellaneous	Hand held vacuum cleaner	4	1	1.9	27
	Cordless electric plug	3	1	1.2	9
	Electric toothbrush	3	1	1.8	12
	Perfume burner	5	1	3.3	3
	Low-voltage halogen lamp	3	3	3.0	1
	Washing machine	7	1	4.0	2
	Electric bed	5	5	5.0	2
	Lightning protection device	1	1	1.0	1
	Baby monitor	3	1	2.0	3
	Water treatment system	7	2	3.2	5

Source : Sidler, 2000.

The average annual household consumption of standby power mode is 235 kWh/year, representing 7 per cent of the total electricity consumption, excluding electric space and water heating, in the sample of households.

The United States case is an example of a bottom-up estimate. Hundreds of individual appliances of all ages were measured in homes, stores and repair shops. The average annual standby energy consumption for each appliance was then multiplied by the number of those appliances found in American homes (which was known from several surveys conducted by the government and private firms). The process was repeated for each appliance found in American homes. The total electricity consumption for all appliances was then divided by the 100 million US homes to give the average standby energy consumption per home.

Similar approaches have been used in Germany, Switzerland and the Netherlands, although fewer types of appliances were included or directly measured. The German and Swiss studies, for example, relied heavily on standby power measurements of new appliances reported in consumer magazines. The standby power consumption of new appliances does not always equal that of older appliances; it can be higher or lower.

Table A2.2 summarises estimates of residential standby power consumption in nine countries. The listed values have been adapted from sources that can be readily compared. Some countries have unique appliances that require standby power. In Japan, the "shower-toilet" consumes 5 watts and rice cookers consume 1 to 2 watts. In France, the Minitel communications systems draw 5 to 9 watts. In Australia, the chargers for electrified fences consume standby power.

Table A2.2 presents standby power levels with three different metrics (to reflect the different types of estimates made by the original researchers). New Zealand and Australia appear to have the highest average standby power consumption. Several factors may account for this. First, the New Zealand estimate is based on a field study of only 29 homes, so the sample may not be representative. Second, it includes the energy used by some appliances that clearly do not fit into the standby definition, notably heated towel rails and several defective refrigerators. Third, New Zealand and Australian appliances operate at 240 volts. Since power supplies have higher losses at higher voltages, the same appliance may have higher standby power consumption in Australia than in Japan (100 volts). Standby power use is also very high in Japan. This is a result of high penetration of electronic appliances and white goods with microprocessor controls, such as air-conditioners, gas water heaters, and washing machines. Switzerland appears to have the lowest standby power use, but this is probably because only the most widely used appliances were included in the estimate.

Table A2.2 **Estimates of residential standby in nine countries**

Country	Average residential standby power use (watts)	Annual electricity use (kWh/yr)	Fraction of total residential electricity use	Source and type of estimate	Notes
Australia	86.8	760	11.6%	(Marker and Harrington, 2001) AGO	Field survey of 64 households.
France	27	235	7%	(Sidler, 2000) Enertech	Based on field measurements in 178 homes. Some appliances with standby modes may have been overlooked.
Germany	44	389	10%	(Rath *et al.*, 1997) Bottom-up	May include standby losses from storage water heaters.
Japan	46	398	9.4%	(Matsunaga, 2001) ECCJ	Based on field measurements in 51 homes.
Netherlands	37	330	10%	(Siderius, 1995) Bottom-up	Based on typical standby power use of major appliances. It does not include less common appliances, so the actual value may be higher.
New Zealand	100	880	11%	(EECA, 1999)	Based on a field study of 29 homes. It includes a few heated towel rails and malfunctioning appliances.
Switzerland	19	170	3%	(Meyer & Schaltegger AG, 1999) Bottom-up	Only includes TVs, VCRs, satellite receivers, stereos, some rechargeable appliances, cordless telephones and PCs.
United Kingdom	32	277	8%	(Vowles 2001) ECI Oxford	Field estimate for 32 households.

(continued)

Country	Average residential standby power use (watts)	Annual electricity use (kWh/yr)	Fraction of total residential electricity use	Source and type of estimate	Notes
United States	50	440	5%	(Rainer *et al.*, 1996) Bottom-up	Based on measurements of individual appliances and then adjusted for the number of each appliance in an average home.

In spite of the uncertainty and known errors in some of these estimates, the absolute levels of standby power use are similar. The range in percentages of residential electricity consumed as standby power is much larger because total electricity consumption in homes varies much more. The general agreement of the two approaches underpins an estimate that standby power use in developed countries ranges from 20 to 60 watts per household.

A GLOBAL ESTIMATE OF STANDBY POWER CONSUMPTION IN THE RESIDENTIAL SECTOR

The goal is to estimate the global consumption of standby power and the CO_2 emissions that result from power generation to meet this demand. The data presented in Table A2.2 describe the residential standby power consumption in several developed countries. Even though the approaches of the studies were different, the fact that these countries have similar appliances and similar appliance penetration levels are good reasons to believe that similar levels of standby power use exist in other developed countries. The amount of standby power consumed in less developed countries is uncertain.

A global estimate of standby power use should, ideally, include estimates for the less developed countries. But the situations in the developed countries are significantly different from those in the less developed countries; moreover, the data from those countries are particularly poor. Even where stocks may be known, significant uncertainties exist. For example, most Chinese consumers unplug their VDRs (video-disk players) and VCRs while not in operation. (This is to protect them against voltage spikes and because VDRs and VCRs are principally used to play disks and tapes, not to record broadcast material.) It is difficult to estimate standby power use when the major appliances are unplugged for large fractions of the day. Connection times may increase in the future, especially as more material is available via broadcast, satellite, or cable.

A first step towards a global estimate would be to estimate standby power use in 29 OECD countries. These countries account for roughly 65 per cent of the world's electricity use and 54 per cent of global CO_2 emissions. Table A2.3 presents a summary of this bottom-up estimation.

For each OECD country, the table begins with the number of households and the estimated average standby power use based on the findings summarised in Table A2.2. Countries with similar economies, such as Canada and the United States, Austria and Germany, Switzerland and France, or New Zealand and Australia, are assumed to have the same average standby power use. For the OECD countries with the lowest GDPs (Czech Republic, Greece, Hungary, Mexico, Poland, Portugal, South Korea, Spain and Turkey), the conservative figure of 20 watts of standby power consumption per household is assumed.

From this information it is then possible to derive an estimate of the total standby electricity consumption in the OECD residential sector. IEA statistics on CO_2 emissions from electricity production allow the emissions associated with total residential standby power use to be estimated.

Table A2.3 summarises the energy demand and related CO_2 emissions from standby power use in the residential sector of OECD Member countries.

Table A2.3. **Assessment of energy demand and CO_2 emissions from standby power in the residential sector of OECD Member countries**

OECD Member countries	Number of households (millions of units)	Average standby power (W/home)	Total standby power demand (MW)	Total standby energy (TWh/yr)	Total national consumption 1997 (TWh/yr)	Standby as % of national electricity	CO_2 emission ratio 1997 (gCO_2/kWh)	National CO_2 emissions 1997	CO_2 from standby power (Mt)	Standby as % of national CO_2
Australia	7.09	87	617	5.4	171	3.2 %	942	306	5.1	1.7 %
Austria	3.38	44	149	1.3	53	2.5 %	237	64	0.3	0.5 %
Belgium	3.85	27	104	0.9	78	1.2 %	339	123	0.3	0.3 %
Canada	11.7	50	585	5.1	514	1.0 %	189	477	1	0.2 %
Czech Republic	3.48	20	70	0.6	58	1.1 %	677	121	1.4	0.2 %
Denmark	2.35	39	92	0.8	35	2.3 %	554	62	0.4	0.7 %
Finland	2.2	39	86	0.8	74	1.0 %	294	64	0.2	0.3 %
France	23.14	27	625	5.5	410	1.3 %	82	363	0.4	0.1 %
Germany	36.03	44	1 585	13.9	527	2.6 %	690	884	9.6	1.1 %
Greece	3.65	20	73	0.6	42	1.5 %	980	81	0.6	0.8 %
Hungary	3.85	20	77	0.7	33	2.0 %	624	58	0.4	0.7 %
Iceland	0.1	39	0	0	5	0.0 %	1	2	0	0.0 %
Ireland	0.87	32	28	0.2	18	1.4 %	875	38	0.2	0.6 %
Italy	22.69	27	613	5.4	273	2.0 %	605	424	3.2	0.8 %

(continued)

OECD Member countries	Number of households (millions of units)	Average standby power (W/home)	Total standby power demand (MW)	Total standby energy (TWh/yr)	Total national consumption 1997 (TWh/yr)	Standby as % of national electricity	CO_2 emission ratio 1997 (gCO_2/kWh)	National CO_2 emissions 1997	CO_2 from standby power (Mtons)	Standby as % of national CO_2
Japan	41.37	46	1 903	16.7	1 001	1.7 %	439	1 173	7.3	0.6 %
Luxembourg	0.2	44	0	0	6	0.0 %	1 000	9	0	0.0 %
Mexico	21.08	20	422	3.7	152	2.4 %	629	346	2.3	0.7 %
Netherlands	6.51	37	241	2.1	96	2.2 %	522	184	1.1	0.6 %
New Zealand	1.26	87	110	1	33	2.9 %	145	33	0.1	0.4 %
Norway	1.93	39	75	0.7	107	0.6 %	3	34	0	0.0 %
Poland	11.8	20	236	2.1	124	1.7 %	921	350	1.9	0.5 %
Portugal	3.66	20	73	0.6	34	1.9 %	499	52	0.3	0.6 %
South Korea	13.99	20	280	2.5	236	1.0 %	411	422	1	0.2 %
Spain	14.94	20	299	2.6	167	1.6 %	408	254	1.1	0.4 %
Sweden	3.97	39	155	1.4	136	1.0 %	35	53	0	0.1 %
Switzerland	2.98	27	80	0.7	52	1.4 %	10	45	0	0.0 %
Turkey	15.09	20	302	2.6	87	3.0 %	685	187	1.8	1.0 %
United Kingdom	21.93	32	702	6.1	337	1.8 %	565	555	3.5	0.6 %
United States	101.04	50	5 052	44.3	3 503	1.3 %	648	5 470	28.7	0.5 %
OECD	**386**	**38**	**14 634**	**128**	**8 362**	**1.5 %**	**530**	**12 235**	**68**	**0.6 %**

With this method, standby electricity in the residential sector of OECD Member countries is estimated to be responsible for 1.5 per cent of total electricity consumption and to contribute 0.6 per cent (68 million tonnes) of CO_2 emissions from the electricity sector. This represents the annual CO_2 emissions of 24 million European-type cars.

The total demand for standby power in the residential sector amounts to 15 GW. For comparison, the installed capacity of wind turbines, worldwide and in the same period, is slightly over 10 GW, with a total electricity production below 30 TWh/year. It would take many years and large capital investments to install enough wind turbines to offset the energy being consumed in the standby mode of the numerous electric devices studied here.

ANNEX 3
CURRENT PRODUCT MARKETS AND FUTURE DEVELOPMENTS

This annex documents the current characteristics of products that consume power in standby mode and likely future developments. This information was used in the discussions on policies and measures to address standby power in Task Force 3 of the IEA's programme on standby power.

Table A3.1 **Consumer electronics: Analysis of the current situation**

Characteristics	Whole product group	Television	VCR	IRD (set-top boxes)	Audio equipment
1. The current standby situation		Mostly simple: remote control function.	Complex, several functions.	Complex, several functions, including remote access. (Depends on market situation.)	Mostly simple: remote control. Hive-systems: + clock. Sometimes "off" mode consumption.
2. Market situation		Mature market; products sold to end consumers.	Mature market (replacement by other products is expected); products sold to consumers.	Starting/ developing market. Most boxes are part of a package from a service provider, so manufacturers do not sell to consumers.	Mature market; products sold to consumers.
3. Importance to buyers of energy efficiency in general or standby power consumption	Low to zero.	Low; issue raised by some consumer organisations and label organisations.	Low; issue raised by some consumer organisations and label organisations.	Zero.	Low; issue raised by some consumer organisations and label organisations.

(continued)

Characteristics	Whole product group	Television	VCR	IRD (set-top boxes)	Audio equipment
4. Competitiveness of the market	Highly competitive.	Price competition; premiums only for top-of-the-market/design products.	Price competition.	Price/functionality competition.	Price competition; premiums only for top-of-the-market/design products.
5. (Main) actors or components that influence standby consumption	In general, the power supply is an important component.	Manufacturers. Main components (CRT, chipsets) manufactured by a few manufacturers.	Manufacturers.	Service providers, IRD manufacturers, chipset manufacturers. Software plays an important role. Chipsets only available from a few manufactures.	Manufacturers.
6. Dynamics of the market	Time to market of new products mostly determined by social acceptance.	Chassis lifetime: five to six years. Major developments: colour TV, wide screen TV, 100 Hz TV, digital TV.	Chassis lifetime: five to six years?	Technical developments determined by new chip generations.	
7. Main drivers for improving efficiency in standby power use	Internal (manufacturer) environmental policy, external policies (labelling, voluntary agreements), autonomous technical developments.				
8. Main barriers to improving efficiency in standby power use	Costs, need to co-ordinate improvements with (re)design cycle.			Co-ordination between several market parties required. Lack of standards for power management.	
9. Differences among the major world markets (US, Japan, EU)	Technical differences are decreasing.	US: no "off" mode.		Differences in market structure.	

Table A3.2 **Consumer electronics: Developments expected in the next five to ten years**

Characteristics	Whole product group	Television	VCR	IRD (set-top boxes)	Audio-equipment
1. Obsolescence, replacement products		Screen might evolve into a separate product (display) that can be connected to several sources.	VCR is likely to be replaced with DVD-recorder, IRD with hard disk or recorder with solid-state storage.		Cassette deck is likely to be replaced with CD recorder/writer or solid state recorder (e.g. MP3).
2. New products on the market (standby characteristics)		Flat screens (might have external power supply: power consumption in "off" mode).			MP3 players (or similar devices).
3. Technological developments	Developments in control circuit (integrated circuit) technology.	Flat screen technology (affects only (?) "on"-mode).	Development of other (than tape) storage media.		Development of storage media.
4. Impact of connection to a (home) network	Connection to (home) network has to be maintained.	Connection to Internet.	Video on demand (via Internet).	IRD could develop into "gateway" to the home (MHP concept).	Connection to Internet.
5. Impact of product integration and/or differentiation		TV screen used for several purposes.		Possible integration with PC.	Integration with TV.
6. Development of market volume		Number of screens per household might increase.		Increasing to (at least) one IRD per household. US: more use of satellite systems as compared with cable systems.	Increasing because of personalised and situationalised audio equipment.

Total standby energy consumption associated with consumer electronics is likely to increase in the residential sector. The "off" mode will become obsolete in televisions and audio equipment. The main drivers for increasing power consumption in standby modes will be the connection of products to (home) networks. These network connections will need to be powered even if they are not used. However, if power management is introduced for networked products, total power consumption in both active and standby modes could decrease.

Table A3.3 **Information technology: Analysis of the current situation**

Characteristics	Whole product group	PC (system unit)	Monitor
1. The current standby situation		Complex: several functions. Power management. New units consume power in "off" mode, or do not have an "off" mode.	Moderate: wait for wake-up signal, but various standby modes to balance power consumption and recovery time. If monitor has USB ports, these need to be powered too.
2. Market situation		Mature market; products are sold to both consumers and businesses.	Mature market; products are sold to both consumers and businesses.
3. Importance to buyers of energy efficiency in general or standby power consumption		Low to moderate; issue raised by label organisations.	Low to moderate; issue raised by label organisations.
4. Competitiveness of the market		Various manufacturers. Competition on price; premium only for state-of-the-art products.	Various manufacturers. Competition on price; premium only for top-of-the-market products.
5. (Main) actors or components that influence standby consumption		Chipset and motherboard manufacturers, software developers. Main components available from a few manufacturers but in many versions.	Power supply manufacturers.
6. Dynamics of the market		Continuous improvements in processing speed, storage capacity, etc. Major functional improvements determined by social acceptance and software base.	Relatively slow. Major developments: colour monitors, "digital" monitors, LCD monitors. (Monitor size has increased over the years: affects "on" mode only.)
7. Main drivers for improving efficiency in standby power use	Internal environmental policy, external policies (labelling), autonomous technical developments.		
8. Main barriers to improving efficiency in standby power use		Costs. Co-ordination with software development. Lack of standardisation on power management.	Costs.
9. Differences among the major world markets (US, Japan, EU)	Very little.		

Table A3.4 **Information technology: Developments expected in the next five to ten years**

Characteristics	Whole product group	PC (system unit)	Monitor
1. Obsolescence, replacement products		Not likely within next five years.	None.
2. New products on the market (standby characteristics)		"Hidden" PCs: Internet terminals, communication switches, etc. Standby characteristics will be equal to those of PCs.	
3. Technological developments		Increasing processing (graphics) and storage capacity.	Further developments in flat screen technology: larger screens.
4. Impact of connection to a (home) network		PC will be continuously connected to one network (or several).	Monitors could be separately connected to the network.
5. Impact of product integration and/or differentiation		Integration with, e.g., IRD (multimedia home platform) is uncertain. Further differentiation of PC: desktop, laptop, palmtop, "hidden" PCs.	Possible integration of computer and television displays.
6. Development of market volume		Increasing volume.	Increasing volume because of situationalised use (e.g., kitchen).

As with consumer electronics, the "off" mode will become obsolete for information technology (IT) appliances such as PCs and monitors. IT equipment will be connected to networks and will require some power in standby modes around the clock. Standby power consumption is consequently very likely to increase, although implementation of power management could decrease total power consumption.

Table A3.5 **Office equipment: Analysis of the current situation**

Characteristics	Whole product group	Printer	Copier	Fax machine
1. The current standby situation		Moderate to complex (network printers).	Complex: various standby modes to balance power consumption and waiting time.	Complex.
2. Market situation		Products are sold to both consumers and businesses. Network printers are sold primarily to businesses, smaller printers to consumers.	Products are (mostly) sold to businesses; copiers are often leased or rented. Relatively low volume.	Declining market (especially consumer market). Products are sold to both consumers and businesses.
3. Importance to buyers of energy efficiency in general or standby power consumption	Low to moderate.	Low to moderate; issue raised by some consumer organisations and label organisations.	Low to moderate; issue raised by some consumer organisations and label organisations.	Low to moderate; issue raised by some consumer organisations and label organisations.
4. Competitiveness of the market	High.	Price competition in low-end market; price/functionality (colour, printing speed) competition in higher-end market.	Copier is often only part of a larger package (including maintenance).	Price and feature competition.
5. (Main) actors or components that influence standby consumption		Power supply manufacturers.	Manufacturers of core components (drum, toner), electronics (including power supply and network connections).	Manufacturers.
6. Dynamics of the market	Printer and copier (and fax machine and scanner) tend to be integrated into one appliance.	New models appear every six months. Major developments: multifunctional devices, network connectivity, colour laser printing.	New models every 12 months. Chassis lifetime: five to six years. Major developments: network connectivity, colour copying, digital, multifunctional devices.	Because of declining market, innovations not expected.

(continued)

Characteristics	Whole product group	Printer	Copier	Fax machine
7. Main drivers for improving efficiency in standby power use	Internal (manufacturer) environmental policy, external policies (labelling), autonomous technical developments.			
8. Main barriers to improving efficiency in standby power use		Costs. Increasing number of printers with external power supplies.	Costs. New developments (network connectivity).	
9. Differences among the major world markets (US, Japan, EU)	Very little.			

Table A3.6 **Office equipment: Developments expected in the next five to ten years**

Characteristics	Whole product group	Printer	Copier	Fax machine
1. Obsolescence, replacement products		Matrix printer will become obsolete; is replaced by ink jet or laser printer.	Black and white copier will be gradually replaced by colour copier. Analog copiers will become obsolete and will be replaced by digital (multifunctional) copier.	Fax machine will become obsolete or at least rare. Availability of electronic signature will stimulate obsolescence. Fax function will be replaced by e-mail.
2. New products on the market (standby characteristics)		(Low price) colour laser printers. Multifunctional devices. Standby characteristics: complex.	(Low price) colour copiers. Multifunctional devices. Standby characteristics: complex.	
3. Technological developments	It is not expected that the "paperless" (or even "less paper") office will be realised within the next ten years.			
4. Impact of connection to a (home) network		Connection with network has to be maintained. No connection with computer needed.	Connection with network has to be maintained.	
5. Impact of product integration and/or differentiation		Multifunctional devices will probably lead to fewer appliances, but with more complex standby power management.	Multifunctional devices will probably lead to fewer appliances, but with more complex standby power management.	Fax function, if needed, will become integrated into multifunctional device.
6. Development of market volume		Consumer market could grow to one printer per household (average). Business market will grow related to volume of office space. Impact of multifunctional devices on volume?	Market will grow related to volume of office space. Impact of multifunctional devices on volume?	Volume will decline.

Office equipment will be increasingly networked, requiring standby power continuously. Colour copying could have an impact on standby power requirements, for example for keeping the drums at the correct temperature.

Table A3.7 **External power supplies and chargers*: Analysis of the current situation**

Characteristics	Whole product group	External power supply (wall pack)	Battery charger
1. The current situation		Simple: only no-load situation is taken into account.	Moderate: both no load, and situation where batteries are fully charged.
2. Market situation		Products are sold primarily to businesses. In sales to consumers, external power supply is usually part of a product.	Products are sold to both consumers and businesses.
3. Importance to buyers of energy efficiency in general or standby power consumption	Zero.	Zero.	Zero.
4. Competitiveness of the market	High.	Several large manufacturers; many small. Competition on price.	
5. (Main) actors or components that influence standby consumption		Control circuit (integrated circuit) designers/ manufacturers.	Control circuit (integrated circuit) designers/ manufacturers.
6. Dynamics of the market		New models every ? months. Major developments: electronic wall packs.	New models every ? months. Developments follow developments in wall packs and batteries.
7. Main drivers for improving efficiency in standby power use		Costs. Switching to electronic designs (and lower no-load losses) results in lower product costs and often lower shipment weight (and cost).	
8. Main barriers to improving efficiency in standby power use		Costs.	For mobile telephones: different interfaces between batteries and charger for each brand/model make exchange between brands/models, and thus fewer chargers, not feasible.
9. Differences among the major world markets (US, Japan, EU)	Very little (e.g. power cord); most products function on mains voltages between 110 and 230 V.		

** Note that batteries in mobile telephones and hand-held kitchen vacuum cleaners can be "charged" by external power supplies (wall packs). In this case, the "charging intelligence" is within the battery, and the power supply is only supplying a (constant) low voltage to the charger.*

Table A3.8 **External power supplies and chargers: Developments expected in the next five to ten years**

Characteristics	Whole product group	External power supply (wall pack)	Battery charger
1. Obsolescence, replacement products	It is not expected that these products will become obsolete in the next five to ten years. More and more portable (and battery-powered) products will require more and more wall packs and battery chargers.		
2. New products on the market (standby characteristics)			
3. Technological developments		Size and weight will decrease. Further development of electronic control circuits for higher power levels.	Shorter charging times, longer battery life.
4. Impact of connection to a (home) network		None.	None.
5. Impact of product integration and/or differentiation		Different power output levels needed?	
6. Development of market volume		Increasing volume (due to booming market for mobile telephones).	

Standby/no-load power consumption may decrease through developments in control circuits for wall packs. Intelligent battery chargers may require more standby power, although this will decrease total consumption.

Table 3.9 **White goods: Analysis of the current situation**

Characteristics	Whole product group	Refrigerator	Washing machine	Microwave oven	Cook top
1. The current standby situation		Simple/complex: standby consumption refers to displays. Standby consumption "integrated" with total consumption.	In most cases, no standby power use when product is switched off; but some power consumed when washing cycle is finished and product is not yet switched off.	Simple to moderate: clock, timer function.	Simple to moderate: clock, timer function.
2. Market situation	Products are sold to consumers. Mostly mature markets.				
3. Importance to buyers of energy efficiency in general or standby power consumption	Focus is on energy efficiency.	Relatively high (energy labelling in several parts of the world).	Relatively high (energy labelling in several parts of the world).	Zero.	Zero.
4. Competitiveness of the market	Several large (worldwide) manufacturers.	Competition on price, energy efficiency.	Competition on price, quality (performance/ durability), energy efficiency.	Competition on price and functionality.	Competition on price and functionality (and in US on type: gas versus electric).
5. (Main) actors or components that influence standby consumption				Power supply and electronic control. Manufacturers/ designers.	Power supply and electronic control. Manufacturers/ designers.
6. Dynamics of the market	New models every year but major developments are occurring much less frequently than before.				

(continued)

Characteristics	Whole product group	Refrigerator	Washing machine	Microwave oven	Cook top
7. Main drivers for improving efficiency in standby power use	If any (NB: refers to standby efficiency): autonomous developments.				
8. Main barriers to improving efficiency in standby power use					More complex cook tops.
9. Differences among the major world markets (US, Japan, EU)	Major differences for washing machines (related to washing "culture/history") and refrigerators.			Only aesthetic differences.	Only aesthetic differences.

Table A3.10 **White goods: Developments expected in the next five to ten years**

Characteristics	Whole product group	Refrigerator	Washing machine	Microwave oven	Cook top
1. Obsolescence, replacement products	It is not expected that these products will become obsolete.				
2. New products on the market (standby characteristics)			Washing machine may become part of a service-related concept in which the consumer does not pay for the machine but for the functionality (pay per washing cycle). Standby situation becomes more complex.		

(continued)

Characteristics	Whole product group	Refrigerator	Washing machine	Microwave oven	Cook top
3. Technological developments	See also 4. Other developments are related to "on" mode.				More cook tops (also gas) will be equipped with clocks, digital displays and electronic controls.
4. Impact of connection to a (home) network	It is expected that (new) major white goods will be connected to a (home) network.				
5. Impact of product integration and/or differentiation	None.				
6. Development of market volume	Development will reflect growth in the number of households.				

Because of network connections, standby power consumption by white goods is expected to increase. Total standby consumption will also increase because all cook tops, including gas cook tops, will be equipped with display/clock features and electronic controls.

ANNEX 4
WEB PAGES USED FOR THIS BOOK

ENERGY LABELS THAT INCLUDE STANDBY

www.energielabel.ch
www.gealabel.org
www.europa.eu.int/comm/environment/ecolabel
www.blauer-engel.de
www.ecolabel.no
www.boilers.org.uk (database transforming the market for boilers)

MARKET TRANSFORMATION ACTIVITIES

www.environment.detr.gov.uk/mtp/index.htm
www.aceee.org
www.neep.org
www.neea.org

STANDBY DATA

http://www.iea.org/standby/index.html
http://eetd.lbl.gov/standby/data.html
www.eia.doe.gov
http://www.energyefficient.com.au/standby/index.html

ENERGY STAR

www.energystar.gov.au – Australia
www.energystar.gov - USA

INTERNATIONAL ORGANISATIONS

www.apecsec.org.sg/
www.climatetech.net/home.shtml
europa.eu.int/comm/role_en.htm
www.wto.org/
www.gealabel.org/
www.iec.ch/
www.iea.org
www.wbcsd.ch/
www.unfccc.de

ANNEX 5
A COMPILATION OF THE WORLD'S LOWEST STANDBY POWER FOR EACH FAMILY OF EQUIPMENT (WORLD'S TOP RUNNERS)

Product group	Product	Make	Model	Watts
Battery-powered devices	Battery charger (NiCd)	GE/Sanyo	NC1280	0.5
	Cell telephone	Motorola	StarTac	0.2
		LG		0.2
	Power tool	Craftsman	14800-08	0.6
	Shaver, hair trimmer	Remington	5BF1-C	0.4
	Toothbrush	Braun	4728	0.9
	Vacuum cleaner (hand-held)	Dirtdevil	BV2000	1.7
Home	Carbon monoxide detector	Enzone	AirZone II	0.0
	Doorbell	China Tech	CH570201	1.1
	Garage door opener	Moore-O-Matic	800-21C	1.4
	Motion sensor	Reflex	SL5511W	0.6
	Security system	Safe House	Radio Shack 49-485	4.5
	Sprinkler controls	Optima	PRT-6	2.1
	Timer	First Alert	94N2	0.6
Kitchen	Breadmaker	Zojirushi	BBCC-V	0.4
	Microwave oven	Kenmore	721.6810079	0.9
	Range	Whirlpool	69676	0.9
	Rice cooker	National	SR-GE10-N	1.5
Office	Modem, analog	US Robotics	LR64979	1.0
	Modem, digital	Com21 Inc.	CP1000D	9.8
	Telephone/fax machine/copier	Panasonic	KX-F580	3.3
	Power speaker	Radio Shack	Pro SW-10P	0.0
	Printer, ink/bubbleJet	Hewlett-Packard	DeskJet 310	4.2
Set-tops	Cable box, analog	Scientific Atlanta	8530338.1	2.4
	Cable box, digital	General Instruments	DCT1000	19.7
	Digital television decoder	Panasonic	TV-DST50W	5.1
	Internet terminal	Philips-Magnavox	MAT960A102	4.4
	Satellite system	RCA	DRD503RBC.1	8.8
	Video game	SEGA	DCX	0.0
Telephony	Answering machine	GE	2-9805A	1.8
	Answering machine/cordless telephone	Uniden	XCAI680	2.1
	Cordless telephone	GE	2-9910B	1.1

(continued)

Product group	Product	Make	Model	Watts
Audio/Video	Television	Nokia	7177	<0.1
	VCR	JVC	HR-J200U	1.5
	DVD player	Sony	DVPS500D	1.3
	Portable stereo	Sony	CFM10	0.5
	Compact system	Panasonic	SAAK27	0.5
	Alarm clock	Radio Shack	63-637	0.7
	Radio, clock	Sony	ICF-25	0.9

ANNEX 6 INTERNATIONAL PROCESSES FOR REDUCING STANDBY LOSSES

INTRODUCTION

This annex summarises international and regional processes that could be used to influence national, industry and consumer activities to address standby power.

Table A6.1 **Summary of relevant international processes**

International organisation	Main processes
APEC	Asia Pacific Economic Co-operation
CTI	Climate Technology Initiative ■ technology assessment, analysis and strategy working group (Co-operative Technology Implementation Plans) ■ technology awards ■ training (e.g. energy efficiency workshops)
EC	European Commission ■ voluntary agreements with industry ■ initiate policy directives at the European level
GATT	General Agreement on Tariffs and Trade ■ Technical Barriers to Trade (TBT) Agreement ■ technical assistance
GEA	Group for Efficient Appliances ■ information campaigns aiming to harmonise procedures for measuring efficiency levels
IEC	International Electrotechnical Commission ■ electrotechnical and IT standardisation ■ technology awards
IEA	International Energy Agency ■ analyses and publications ■ exchange of information ■ co-operative agreements covering energy efficiency, market transformation activities (information exchange, research studies, specific projects)

(continued)

International organisation	Main processes
WBCSD	World Business Council for Sustainable Development ■ contributing a business perspective to policy development through research; research on resource efficiency includes development of indicators and monitoring and reporting guidelines
UNFCCC	United Nations Framework Convention on Climate Change

ASIA PACIFIC ECONOMIC CO-OPERATION

http://www.apecsec.org.sg/

APEC's "Action Programme for Energy" is working towards common action on standards and test protocols. One of the main reasons for this work is to achieve trade benefits for the Asia Pacific region: "Major trade and economic gains will arise from: clarifying production and marketing requirements in all member economies against recognised and agreed benchmarks; increasing certainty among market suppliers in terms of production planning; agreeing test protocols, with the potential for reduced testing and retesting requirements (and hence costs); and increasing certainty among regulators on accreditation procedures and quality assurance processes, further reducing technical and administrative requirements and costs."

APEC's Energy Efficiency and Conservation Expert Group is working on "Acceptance of equivalence in accreditation and increasing harmonisation of energy standards", with the aim of reducing costs to both governments and businesses by both acceptance of equivalence in accreditation and closer harmonisation of standards for energy products, appliances and services where cost-effective. APEC believes that major trade and economic gains will be made if energy standards can be harmonised, or common protocols agreed. Major benefits will derive from the establishment of a harmonised network of protocols in member economies. These will cover areas such as basic product performance, methods of testing for determining energy consumption, accreditation systems, processes for laboratories, and quality-assurance systems and procedures.

CLIMATE TECHNOLOGY INITIATIVE (CTI)

http://www.climatetech.net/home.shtml

The CTI is a voluntary co-operative effort among OECD governments that aims to foster international co-operation for the accelerated development and diffusion of climate-friendly technologies and practices. The secretariat for the CTI is managed by the IEA. The CTI seeks to collaborate closely with developing countries and economies in transition, to work in partnership with stakeholders, and to mobilise expertise on a voluntary basis. Three working groups have been set up under the CTI:

- Working group on capacity building
- Working group on research and development
- Working group on technology assessment, analysis and strategy.

The CTI technology awards, which are given annually, recognise success in stimulating the adoption of climate-friendly technologies and practices in developing countries and in economies in transition. CTI training courses have included a workshop on energy efficiency that sets out energy efficiency policies, programmes and trends.

The CTI intends to continue co-ordinating CTI/industry joint seminars, expand the existing technology information centres and better identify their needs for further support and assistance, enhance training and capacity building for development of technology measures, and strengthen ties with the IEA Greenhouse Gas Technology Information Exchange (GREENTIE).

As well as governments, current CTI partners are drawn from a variety of sectors and interests including the E-7 group of electric utilities, the International Organization for Standardization (ISO), the United Nations Environment Programme (UNEP), the Versailles Agreement on Advanced Materials and Standards (VAMAS) and the World Business Council for Sustainable Development (WBCSD)[29].

EUROPEAN COMMISSION (EC)

http://europa.eu.int/comm/role_en.htm

The European Commission develops energy policy initiatives and makes proposals on legislation for consideration by member States in the Council of the European Union. As the executive body of the Union responsible for implementing and managing policy, the European Commission is a key player in the development of policies to address standby power.

The Commission works in close partnership with the other European institutions and with the governments of the member States. The Commission also acts as the guardian of the EU treaties to ensure that EU legislation is applied correctly by the member States. It can institute legal proceedings against member States or businesses that fail to comply with European law and, as a last resort, bring them before the European Court of Justice. The Commission also negotiates trade and co-operation agreements with outside countries and groups of countries on behalf of the Union. The Commission has been a very active participant in the IEA programme on standby power.

29. *"Greentimes" article, Michael Rucker, CTI Secretariat, International Energy Agency, November 1998.*

Member States are responsible for implementing policies in the European Union, but the Commission has made use of voluntary approaches to work directly with the consumer electronics industry to improve energy efficiency, including the efficiency of standby modes. For example, the Commission has worked with the European manufacturers of consumer electronics to agree on several codes of conduct on the energy efficiency and the standby power consumption of audio, video and related equipment, wall packs (external power supplies) and digital television service systems. These codes of conduct have been agreed in principle between the Commission and some industry representatives.

The European Commission is working with the US Environmental Protection Agency (EPA) to finalise negotiations on a co-ordinated energy labelling programme for office equipment. Under this agreement, the United States and the European Union will use a common set of energy efficiency specifications and a common logo (the Energy Star logo).

GENERAL AGREEMENT ON TARIFFS AND TRADE (GATT)

http://www.wto.org/

The Technical Barriers to Trade (TBT) Agreement of the GATT emphasises the importance of international standards and the participation of developing countries in international standards. The TBT Agreement contains provisions on technical assistance to developing countries in preparing technical regulations; establishing standards institutions and participating in international standardisation bodies; and establishing regulations, certification bodies, and legal frameworks needed to make membership or participation in international or regional agreements possible.

GROUP FOR EFFICIENT APPLIANCES (GEA)

http://www.gealabel.org/

The GEA is a group of national government bodies, energy agencies and other organisations in Europe aiming to establish a uniform European-wide label for energy-efficient electrical appliances. Members are: Austria, Denmark, Finland, France, Germany, the Netherlands, Sweden and Switzerland.

Each GEA member implements information campaigns appropriate to its consumer market. The participants exchange information on current and planned activities in order to improve co-ordination in the early stage of planned projects. The energy label is not compulsory, although harmonisation of common procedures for measuring efficiency levels is encouraged. The GEA criteria are revised regularly in co-operation with industry. Test methods are harmonised as much as possible with other labelling schemes, such as Energy Star. Members are expected to initiate internationally harmonised activities to meet national or international targets. Harmonisation agreements with other parties require approval by all participants.

Participation is open to all European Energy Network members. Other agencies or organisations can participate only after plenary balloting. The working group chairman organises the exchange of data and information among all signatories. Information is disseminated to all interested parties, which include manufacturers, importers and the European Commission.

INTERNATIONAL ELECTROTECHNICAL COMMISSION (IEC)

http://www.iec.ch/

The IEC's mission is to promote, through its members, international co-operation on all questions of electrotechnical standardisation and related matters, such as the assessment of conformity to standards. Its primary activity is developing and publishing international standards and technical reports. The international standards serve as a basis for national standardisation and as references in the drafting of international tenders and contracts.

Membership consists of more than 50 countries, including a growing number of developing countries. Membership allows countries to participate fully in international standardisation activities. Associate members, often countries of limited resources, have observer status and can participate in all IEC meetings. They have no voting rights. A pre-associate membership status supports countries in forming a national electrotechnical committee with the aim of becoming associate members within five years.

Some 200 technical committees and sub-committees and some 700 working groups carry out the standards work of the IEC. Technical committees are made up of members from national committees. Technical committees prepare documents which are then submitted to the national committees. A technical committee may set up expert project teams. The technical committees decide for themselves when to meet. A proposal for new work generally originates from industry via a national committee. It is then communicated to the members of the appropriate technical committee or sub-committee.

The IEC is a key player in the preparation of international standards in IT through the Joint Technical Committee on Information

Technology (JTC 1), which was formed by an agreement between the IEC and the International Organization for Standardization (ISO).

Another of the IEC's major partners is the World Trade Organization (WTO), through an agreement which recognises that developing international standards can play a critical role in avoiding technical barriers to trade.

The Lord Kelvin Award is given each year to recognise outstanding contributions to global eletrotechnical standardisation over a number of years.

INTERNATIONAL ENERGY AGENCY (IEA)

www.iea.org

Activities of the IEA include:

- Analysis
- Technology policy development
- Sharing of practical experience and exchange of information
- Collaborative research and development and information dissemination
- Publications.

Annex III of the IEA Implementing Agreement[30] on Demand-Side Management focuses on market transformation activities that aim to create buyer demand, and competitive supply, of energy-efficient projects. Standby power consumption is not yet considered explicitly in Annex III but discussions are under way to establish a

30. "Implementing Agreements" are collaborative activities in technology R&D, technology analysis and information dissemination among IEA Member countries and some other countries. "Annexes" are specific tasks under an agreement.

sub-task on standby power within this annex[31]. Aggregated procurement strategies as well as improved buyer awareness through energy rating and labelling are two of the important themes of the market transformation work. Increased involvement of retailers in the promotion of more efficient products will be another major activity.

Work under Annex III:

- Information exchange on energy efficiency market transformation. This might include activities such as producing newsletters and electronic bulletin boards; promoting networking among individuals and organisations; hosting *ad hoc* working groups, workshops or conferences and co-sponsoring a journal on market transformation.
- Research studies on market transformation policy and technology issues. This might include evaluation reports or case studies on market transformation policy, comparative studies of efficiency labelling and quality labels, and comparisons of energy testing and rating methods.
- Co-operative market transformation projects. This includes co-ordinated volume purchasing of energy-efficient products by large-volume government, institutional or corporate purchasers as well as data gathering to develop co-ordinated efficiency labels and quality marks.

31. Personal communication, Verney Ryan, Energy Division, The Building Research Establishment (operating agents for the annex), United Kingdom.

WORLD BUSINESS COUNCIL FOR SUSTAINABLE DEVELOPMENT (WBSCD)

http://www.wbcsd.ch/

The WBCSD is a coalition of 125 international companies that aims to develop closer co-operation among businesses, government and all other organisations concerned with the environment and sustainable development. The council aims to participate in policy development and to share information on best practice among members.

Eco-efficiency is a central tenet of WBCSD's philosophy and is being implemented within member companies in various ways:

- Eco-efficient processes: making resource savings allows companies to decrease the cost of production.
- Reusing by-products: the wastes and by-products of one company are used as alternative resources in another company. Synergies on by-products allow additional improvements towards zero waste.
- Creating new and better products: ecological design rules are followed.
- Eco-efficient markets: companies can further reduce ecological damage by "greening" their supply chain.

The WBCSD has published guidelines for indicators and on best practice in monitoring and reporting eco-efficiency.

Proposals are brought to the relevant WBCSD committee. In the case of standby losses, this could be the Resource Efficiency working group or the Energy and Climate Change working group. If there is sufficient interest, seed capital is sought from companies to enable research to be carried out.

UNITED NATIONS FRAMEWORK CONVENTION ON CLIMATE CHANGE

www.unfccc.de

UNFCCC negotiating meetings provide a good opportunity to raise awareness of standby losses. Measures to reduce standby losses will help IEA countries in meeting their Kyoto commitments. The CTI, the WBCSD and many other groups meet on the margins of UNFCCC meetings and publicise their messages to delegates and observers.

LIST OF ACRONYMS AND ABBREVIATIONS

ACEEE	American Council for an Energy Efficient Economy
ADEME	Agence de l'environnement et de la maîtrise de l'énergie (National Energy and Environment Agency)
AFNOR	Association française de normalisation (French national standards body)
AHAM	Association of Home Appliance Manufacturers
APEC	Asia Pacific Economic Co-operation
ASEW	Association of Municipal Public Utilities for Energy Efficiency (Germany)
ASHRAE	American Society of Heating, Refrigerating and Air-Conditioning Engineers
ASME	American Association of Mechanical Engineers
BATNEEC	best available techniques not entailing excessive costs
BSI	British Standards Institution
CTI	Climate Technology Initiative
CRT	cathode ray tube
DVD	digital versatile disk
EACEM	European Association of Consumer Electronics Manufacturers
EICTA	European Information and Communication Technology Industry Association
EPA	Environmental Protection Agency (United States)
FEA	trade association of enterprises in the household appliance market (Switzerland)
GEA	Group for Efficient Appliances

GED	Gemeinschaft Energielabel Deutschland (German Energy Label Association)
GW	gigawatt, or 1 watt x 10^9
ICT	information and communication technology
IEC	International Electrotechnical Commission
IRD	integrated receiver decoder
ISO	International Organization for Standardization
JIS	Japanese Industrial Standards
kWh/yr	kilowatt-hour/year
LBNL	Lawrence Berkeley National Laboratory
LCD	liquid crystal display
LDO	low dropout
LED	light-emitting diode
METI	Ministry of Economy, Trade and Industry (Japan)
MHP	multimedia home platform
Mt	million tonnes
MW	megawatt of electricity, or 1 watt x 10^6
NOVEM	The Netherlands Agency for Energy and the Environment
PSU	power supply unit
TW	terawatt, or 1 watt x 10^{12}
TWh	terawatt x one hour
UKMTP	The UK Market Transformation Programme
UNFCCC	United Nations Framework Convention on Climate Change
VA	voluntary agreement
VCR	video-cassette recorder
WBCSD	World Business Council for Sustainable Development
WTO	World Trade Organization

REFERENCES

EECA, 1999. *Energy Use in New Zealand Households, Report on the Year Three Analysis of the Household Energy End Use Project (HEEP).* Wellington (New Zealand): Energy Efficiency and Conservation Authority.

Euromonitor, 1999. *World Consumer Markets 1999/2000*, 5th edition. London: Euromonitor International.

Harrington, L., 2000. *Study of Greenhouse Gas Emissions from the Australian Residential Building Sector to 2010.* Canberra (Australia): prepared by Energy Efficiency Strategies, Inc.

International Energy Agency (IEA), 2000. Task Force 1: Definition of Standby Power. Http://www.iea.org/energy/ee.htm. Date accessed: 21 May 2000.

Lebot, B., A. Meier and A. Anglade, 2000. "Global Implication of Standby Power Use". In *Proceedings of ACEEE Summer Study on Energy Efficiency in Buildings.* Pacific Grove, CA: American Council for an Energy-Efficient Economy.

Marker, T. and L. Harrington, 2001. "Standby Power Consumption in Australia: Results of Intrusive Monitoring". Presented at the 3rd International Workshop on Standby Power. IEA, February 2001.

Matsunaga, T., 2001. "Survey of Actual Standby Power Consumption in Japanese Households". Energy Conservation Centre, Japan. Presented at the 3rd International Workshop on Standby Power. IEA, February 2001.

Meier, A., K. Rosen and W. Huber, 1998. "Reducing Leaking Electricity to 1 Watt". In *Proceedings of ACEEE Summer Study on Energy Efficiency in Buildings.* Pacific Grove, California: American Council for an Energy-Efficient Economy.

Meyer & Schaltegger AG, 1999. *Bestimmung des Energieverbrauchs von Unterhaltungselectronikgeraeten, Buerogeraeten und Automaten in der Schweiz*. St Gallen (Switzerland): Meyer & Schaltegger AG.

Nakagami, H., A. Tanaka, C. Murakoshi and B. Litt, 1997. "Standby Electricity Consumption in Japanese Houses" In *Proceedings of First International Conference on Energy Efficiency in Household Appliances*. Florence (Italy): Association of Italian Energy Economics.

National Appliance and Equipment Energy Efficiency Committee, 2000. *Standby Power Consumption: Developing a National Strategy*. Canberra (Australia): Australian Greenhouse Office.

Rainer, L., A. Meier and S. Greenberg, 1996. "Leaking Electricity in Homes". In *Proceedings of ACEEE Summer Study on Energy Efficiency in Buildings*. Pacific Grove, California: American Council for an Energy-Efficient Economy.

Rath, U., M. Hartmann, A. Praeffke and C. Mordziol, 1997. *Klimaschutz durch Minderung von Leerlaufverlusten bei Elektrogeräten*. Forschungsbericht 20408541 UBA-FB 97-071. Berlin: Umweltbundesamt.

Rosen, K. and A. Meier, 2000. "Power Measurements and National Energy Consumption of Televisions and Videocassette Recorders in the USA". *Energy* 25: 219-232.

Siderius, H.-P., 1995. *Household Consumption of Electricity in the Netherlands*. Delft (Netherlands): Van Holsteijn en Kemna.

Sidler, O., 2000. *Campagne de mesures sur le fonctionnement en veille des appareils domestiques*. Report No. 99.07.092. Sophia-Antipolis (France): ADEME.

United States Environmental Protection Agency (US EPA), 2000. Energy Star – Labeled Home Electronics. Http://www.epa.gov/appstar/home_electronics /index.html. Date accessed: 21 May 2000.

Vowles, J., 2001. "Standby Power in UK: Results of an Extensive Survey". Environmental Change Institute, University of Oxford, UK. Presented at the 3rd International Workshop on Standby Power. IEA, February 2001.

Order Form

INTERNATIONAL ENERGY AGENCY

ORGANISATION FOR ECONOMIC CO-OPERATION AND DEVELOPMENT

OECD PARIS CENTRE
Tel: (+33-01) 45 24 81 67
Fax: (+33-01) 49 10 42 76
E-mail: distribution@oecd.org

OECD BONN CENTRE
Tel: (+49-228) 959 12 15
Fax: (+49-228) 959 12 18
E-mail: bonn.contact@oecd.org

OECD MEXICO CENTRE
Tel: (+52-5) 280 12 09
Fax: (+52-5) 280 04 80
E-mail: mexico.contact@oecd.org

Please send your order by mail, fax, or e-mail to your nearest IEA sales point or through the online service: www.oecd.org/bookshop

OECD TOKYO CENTRE
Tel: (+81-3) 3586 2016
Fax: (+81-3) 3584 7929
E-mail: center@oecdtokyo.org

OECD WASHINGTON CENTER
Tel: (+1-202) 785-6323
Toll-free number for orders:
(+1-800) 456-6323
Fax: (+1-202) 785-0350
E-mail: washington.contact@oecd.org

I would like to order the following publications

PUBLICATIONS	ISBN	QTY	PRICE	TOTAL
☐ **Things that Go Blip in the Night**	**92-64-18557-7**		**$100**	
☐ Energy Labels and Standards	92-64-17691-8		$100	
☐ World Energy Outlook 2000	92-64-18513-5		$150	
☐ Dealing with Climate Change	92-64-18560-7		$100	
☐ The Road from Kyoto – Current CO_2 and Transport Policies in the IEA	92-64-18561-5		$75	
☐ Energy Technology and Climate Change – A Call to Action	92-64-18563-1		$75	
☐ Emission Baselines: Estimating the Unknown	92-64-18543-7		$100	
☐ Energy Policies of IEA Countries – 2000 Review (Compendium)	92-64-18565-8		$120	
			TOTAL	

DELIVERY DETAILS

Name　　　　Organisation
Address

Country　　　　Postcode
Telephone　　　　Fax

PAYMENT DETAILS

☐ I enclose a cheque payable to IEA Publications for the sum of US$ ________ or FF ________

☐ Please debit my credit card (tick choice). ☐ Access/Mastercard ☐ Diners ☐ VISA ☐ AMEX

Card no:
Expiry date:
Signature:

IEA PUBLICATIONS, 9, rue de la Fédération, 75739 PARIS Cedex 15
Pre-press by Linéale Production. Printed in France by Sagim
(61 01 06 1 P) ISBN 92-64-18557-7 2001

Cover illustration by Bill Marshall